Intelligent System Applications in Power Engineering

Intelligent System Applications in Power Engineering

EVOLUTIONARY PROGRAMMING
and
NEURAL NETWORKS

Loi Lei Lai
City University, London, UK

JOHN WILEY & SONS

Chichester · New York · Weinheim · Brisbane · Singapore · Toronto

Library of Congress Cataloging-in-Publication Data

Lai, Loi, Lei.
    Intelligent system applications in power engineering :
evolutionary programming and neural networks / Loi Lei Lai
      p. cm.
    Includes bibliographical references and index.
    ISBN 0-471-98095-1 (alk. paper)
    1. Evolutionary programming (Computer science)   2. Neural networks
(Computer science)   3. Power (Mechanics)—Data processing.
    I. Title.
    QA76.618.L35   1998
005.1—dc21   98–6560
                                                CIP

British Library Cataloguing in Publication Data

A catalogue record for this book is available from the British Library

ISBN 0 471 98095 1

Produced from camera-ready copy supplied by the author.
Printed and bound by Antony Rowe Ltd, Eastbourne
This book is printed on acid-free paper responsibly manufactured from sustainable forestry in which at least two trees are planted for each one used for paper production.

To my dear mother Lan and father Nam.

To my family: Li Rong, Qi Hong, Chun Sing and Qi Ling.

CONTENTS

PREFACE

An artificial neural network is an analysis paradigm that is roughly modelled after the massively parallel structure of the brain. It simulates a highly interconnected, parallel computational structure with many relatively simple individual processing elements. Neural networks are seen as techniques for information gathering and utilisation.

Evolutionary computation comprises machine learning optimisation and classification paradigms based on mechanisms of evolution such as biological genetics and natural selection. The evolutionary algorithms include genetic algorithms, evolutionary programming, genetic programming, and evolution strategies.

Many problems are not amenable to solution by classic optimisation techniques for a variety of reasons, for example the objective function of concern does not provide gradient information. Evolutionary computation does not require gradient or higher order information regarding the response surface associated with the objective function, nor is it restricted by non-linear constraints, infeasible regions, stochastic variation, or similar features. Together with neural networks, evolutionary computation provides the basis for an intelligent system.

Since the 1980s, numerous research and development projects have focused on the application of intelligent system techniques to electric power engineering problems. In the fast changing power industry environment of today, intelligent system applications are finding exciting new opportunities for power engineering applications.

This book presents a coherent account of applications of evolutionary programming and neural networks in power engineering. It uses numerous figures to illustrate important ideas and to aid in the comprehension of the principles involved. The book is designed for graduate students in electric power engineering as a textbook for advanced courses. It also serves as a reference book for engineers, computer application specialists, and project managers at utilities, consulting firms, and vendors who wish to understand the use of evolutionary programming and neural networks in power engineering applications.

The object-oriented technique has assisted in the modelling and development of an intelligent system. The system can be readily implemented with the mechanisms offered by object orientation such as software reusability, inheritance, encapsulation, modularity and thus can be easily modified or expanded. The first three chapters deal with neural networks, evolution computing and its hybridisation. The next eight chapters deal with the application of evolutionary programming in power engineering problems.

Economic and environmental pressures require electric utilities to use available generation and transmission more intensively. Reactive power dispatch are cost effective, short lead time methods to increase transmission capacity. Chapters 4 and 5 deal with reactive power planning and dispatch. As transmission systems mature, voltage instability often limits power transfers. It is essential to plan the transmission system accordingly. Chapter 6 deals with transmission network planning. Also to study the performance of a power system, it is essential to have a good knowledge of the generator parameters. Chapter 7 deals with generator parameter estimation.

One of the main aims of privatisation is to promote competition and to allow 'market forces' to play their part in keeping down prices and service quality up. Through more efficient operation and utilisation of electricity, there is room for more activities which would have the beneficial effect of stimulating the economy. High voltage power electronics offers new options in power system planning and operation. For example, flexible AC transmission system (FACTS) improves system dynamic performance, improves parallel flow control and optimises loading for loss minimisation. Chapters 8, 9 and 10 deal with economic dispatch, power flow control in FACTS and co-generation scheduling. Privatisation also places emphasis on reducing outage time; Chapter 11 deals with fault section estimation.

Chapters 12-16 deal with the application of neural networks in power systems. Chapters 12 and 13 are concerned with fault diagnosis in high voltage direct current (HVDC) systems and transformers, respectively. As supply quality is an important issue, Chapter 14 deals with power frequency and harmonic evaluation. When a disturbance, e.g. a fault, occurs in a power system, it is essential to protect the system and estimate the critical clearing time. Chapters 15 and 16 deal with distance relaying and transient stability assessment, respectively.

It is becoming less common to read about an application that uses just neural networks, or just evolutionary computation, or just fuzzy logic. There are many possibilities for combining the above-mentioned methodologies with knowledge elements, and it would be impossible to completely discuss all of them here. The hybrid tools being developed using combinations of neural networks, evolutionary computation and fuzzy logic, combined with knowledge elements, are solving increasingly difficult problems. Examples and references are used so that the reader can grasp the principles without difficulty. Chapter 17 deals with short-term load forecasting with evolutionary programming and neural networks. The book is particularly appropriate at a time of increasing interest in the potential offered by these new computational techniques.

This book presents the state-of-the-art of the methods and procedures necessary for designing and developing an intelligent system. It takes into consideration that theoretical investigation must be extended by practical considerations, especially with respect to the mutual dependencies between intelligent techniques. All important aspects of power engineering are covered in this book. In the introduction to each chapter the importance of the particular area covered in that chapter is indicated. This is designed to stimulate the reader's interest in that area and provide motivation to read about it. Following the introduction, a need for intelligent systems will be provided. This is followed by analysis, problem formulation applications and so forth. This book is written in such a way that the user can select topics without losing continuity.

The book reports cutting-edge research and promising results, and also it provides a solid treatment of evolutionary programming (EP) and neural networks (NNs) to power engineering optimisation problems. It shows from the applications covered in the book, with the type of coding strategy, evolution defined for genetic algorithms (GAs), and particular formulation adopted for the problem, that EP is better than GA, which is an optimisation method used by many researchers already. It gives the reader an insight into the potential of using EP as an optimisation technique. So far, no book exists on the market to introduce the application of EP to power engineering optimisation problems. This technique will be able to benefit power utilities in terms of planning, control, operation, protection and cost saving. At the end of the book, a select bibliography has been provided for further information.

ACKNOWLEDGMENTS

The author wishes to thank Mr Peter Mitchell of Wiley and Dr Brian Cory of the Imperial College of Science, Technology and Medicine, UK for recommending the project.

The author also wishes to thank his researchers for contributing technical input to the book. In alphabetical order, they are Dr A F Adbulbary, Mr H Braun, Mr W L Chan, Dr F N Che, Mr B Gwyn, Dr K K Li, Dr J T Ma, Mr N Rajkumar, Mr A Sichanie and Mr H Subasinghe.

The author also expresses his thanks to the large number of friends who have improved the content of the book by commenting on some sections and chapters. Among them, Professor Kwang Y Lee of Pennsylvania State University, USA, Professor Chen-Ching Liu of University of Washington, USA, Professor Stephen K L Lo of University of Strathclyde, UK, Professor John Macqueen of The National Grid Company plc, UK, Professor Om P Malik of University of Calgary, Canada, Professor Vijay K Sood of Hydro-Quebec, Canada, Dr Hideo Tanaka of Tokyo Electric Power Company, Japan, and Professor Kit Po Wong of University of Western Australia.

My thanks also to Mr H Subasinghe and Mr E Vaseekar for producing the book in the necessary format.

Last but not least, the author thanks Wiley for supporting the preparation of this book and for the extremely pleasant co-operation.

ABOUT THE AUTHOR

Loi Lei Lai obtained his BSc (First Class Honours) and PhD from the University of Aston in Birmingham, UK. He joined Staffordshire Polytechnic (now known as Staffordshire University), UK as a Senior Lecturer in 1984. In 1987, he was an Industrial Fellow to both the GEC Alsthom Turbine Generators Ltd and GEC Alsthom Engineering Research Centre. He took up an academic post at City University in 1989. Currently, he is Head of the Energy Systems Group. Since 1994, he is also a Professor at North China Electric Power University, Beijing, China. In 1995, he was a Research Professor at Tokyo Metropolitan University, Japan. He has been invited to give several plenary sessions in major international conferences organised by the IEEE and IEE. He has also been a member of the international advisory committee for a number of IEE or IEEE International Conferences. He travels regularly to give seminars, short courses and consultancy worldwide. The trips are sponsored by various organisations, such as the United Nations.

1

Object-Oriented Analysis, Design and Development of Artificial Neural Networks

This chapter summarises some basic concepts of artificial neural network (ANN) computing. The ideas of supervised and unsupervised learning are explained. Several different types of ANNs are presented. Object modelling technique is introduced and it is used to provide a clear and unambiguous classification of the ANN architectures and operation of ANNs.

1.1   INTRODUCTION

The theory of artificial neural networks (ANNs) is mainly motivated by the application of neural concepts for innovative technical problem solving. The power of ANNs lies in their ability to find a general solution to a given problem. An ANN may solve a task similar to that solved by the biological brain, but will not be an imitation, neither in material nor means nor constraints. ANNs are massively parallel networks of simple processing elements designed to emulate the functions and structure of the brain. They combine properties such as fault tolerance and robustness to solve very complex problems.

ANNs have their roots in the theory of function approximation and pattern classification. ANN models are characterised by a variety of factors. This determines the type of network and the range of applications in which they can be successfully used. The object modelling technique (OMT) [1] can be used to provide a clear and unambiguous classification of the ANN architectures and to describe the static structure and operation of ANNs.

1.1.1 ANN Architectures

Because of the proliferation in the development of ANN models, the need for an intelligent classification of ANNs has become unavoidable. There are too many difficulties inherent in any attempt to classify ANNs. The object modelling technique (OMT) provides a comprehensive notation that makes it easier to visualise both the structure and the functions of complex ANN software systems. This is also used to help a better visualisation of the object-oriented systems. Among the different number and types of ANN structures, feedforward multilayer perceptron (MLP) ANNs have proven to be the most popular used so far in real industrial problems. This is partly because they perform better over a larger set of different problems, which has been proved by a number of research papers.

1.1.2 Feedforward Multilayer ANNs

In feedforward networks, the neurons are arranged in a feedforward manner, usually in the form of layers. Each neuron may receive inputs from the external environment or from other neurons in preceding layers. Neurons are not allowed to have feedback connections or connections from neurons within the same layer. Feedforward ANNs compute an output pattern in response to a given input pattern. Examples of feedforward ANNs include multilayer perceptron networks (MLP), temporal dynamic ANNs (TDNN) and Hamming networks. Figure 1.1 shows a block diagram of a MLP feedforward neural network.

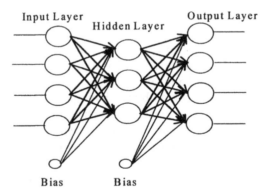

Figure 1.1 Block diagram of MLP feedforward ANN

1.1.3 Structure of ANNs

Every ANN is characterised by a set of coupled differential equations that describe the dynamics of the ANNs [2]. The generalised ANN state equation is

$$\frac{dx}{dt} = F(x, W, U) \tag{1}$$

and the general purpose learning equation is

$$\frac{d\mathbf{W}}{dt} = \mathbf{G}(\mathbf{x}, \mathbf{W}, \mathbf{U}) \tag{2}$$

where
$\mathbf{x} = [x_1(t), x_2(t), \ldots, x_n(t)]^T$ is the activation state space vector,
$\mathbf{W} = [w_{ij}(t)]_{n \times n}$ is the weight matrix and

$\mathbf{U} = [u_1, u_2, \ldots, u_n]^T$ is the time independent external input.

This system of dynamical equations is guaranteed to converge by an **energy** function $E = E(\mathbf{x}, \mathbf{W}, \mathbf{U})$ defined on the state space which is monotonically decreasing.

Physically, an ANN consists of layers of artificial neurons connected by synaptic weights. The basic artificial neuron is modelled as a multi-input, non-linear processing element with weighted interconnections w_{ij}. The neuron processes its input to produce an output according to the following equation:

$$y_j = \Psi \left(\sum_{i=1}^{n} w_{ji} x_i + \theta_j \right) \tag{3}$$

Where
Ψ is the activation function,
θ_j is the bias or offset,
x_i are the inputs $(i = 1, 2, \ldots, n)$, n is the number of inputs,
w_{ij} are the synaptic weights and
y_j are the outputs.

The above equation is often written in a more precise manner by treating the bias as a weight connected to an input that is permanently set to 1:

$$y_j = \Psi \left(\sum_{i=0}^{n} w_{ji} x_i \right) \tag{4}$$

where $w_{j0} = \theta_j$ and $x_0 = 1$.

A schematic diagram of an artificial neuron is shown in Figure 1.2.

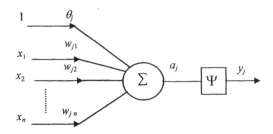

Figure 1.2 Schematic diagram of a simple neuron

The neuron activation function determines the kind of information that one neuron can signal to another. Activation functions may be linear or non-linear. Linear activation functions can be used to provide an approximation of the operations of non-linear models. Non-linear activation functions are used to generate variable and complex performances in ANNs. Sigmoidal functions are normally used as neuron activation functions. However, any function that is continuous, differentiable, monotone increasing and step-like can be used. Two examples of sigmoidal neuron activation functions are shown in Figures 1.3 and 1.4.

Logistic function:

$$y = f(x) = \frac{1}{1 + e^{-kx}} \tag{5}$$

where k is a number.

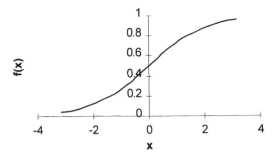

Figure 1.3 Logistic activation function

Hyperbolic tangent:

$$y = f(x) = \frac{e^{kx} - e^{-kx}}{e^{kx} + e^{-kx}} = \tanh(kx) \tag{6}$$

Tanh activation function

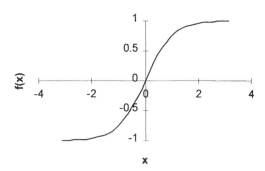

Figure 1.4   Hyperbolic tangent activation function

Neurons with sigmoidal activation functions produce real valued outputs that give the ANN its ability to construct complicated decision boundaries in an n-dimensional feature space. This is important because the smoothness of the generalisation function produced by the neurons, and hence its classification ability, is directly dependent on the nature of the decision boundaries. Another popular activation function is the Gaussian activation function [3]. Figure 1.5 shows a sample Gaussian activation function with zero mean and a unity standard deviation. The local properties of the Gaussian activation function make them more desirable than sigmoidal activation functions in pattern recognition problems. This is due to the fact that Gaussian neurons are able to produce sharper decision boundaries than their sigmoidal counterparts.

The input vectors, $x \in \Re^{n}$, presented to the ANN can be binary data or scaled/normalised real data ranging between -1.0 and +1.0 or 0.0 and +1.0 depending on the nature of the neuron activation function. Most neural learning algorithms have no self-normalising abilities, and hence scaling or normalisation is necessary to prevent saturation of the neuron activation function. In some cases, such as unsupervised learning, the algorithms actually require that the input data and the initial network weights are normalised before learning can take place. Ordinarily, the neuron weights are initialised to small random numbers when the ANN is first created.

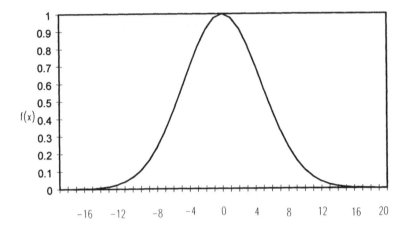

Figure 1.5 Guassian activation function

The structure and operation of a general ANN can be described in graphical terms using models in the object modelling technique. Before doing that, let us examine the object-oriented approach briefly.

1.2   OBJECT-ORIENTED DESIGN (OOD)

1.2.1 Object-Oriented Analysis

Object-oriented software is organised as a collection of distinct objects that co-operate through message-passing to solve a given problem. Each object is an abstraction that encapsulates both data and the operations that act on the data. The analysis process consists of the following activities:

- finding objects;
- organising objects;
- describing object interaction;
- defining operations on objects;
- object-oriented design;
- implementation;
- object-oriented testing.

We look at each in turn.

1.2.2 Finding Objects

Objects can be found as naturally occurring entities or items in the problem domain. An object is typically a noun which exists in the problem statement or problem domain. A domain and/or behaviour analysis can be carried out to help uncover relevant objects in a particular application domain. Domain analysis is an attempt to identify objects, operations and relationships that domain experts perceive to be important about a particular problem or application domain [4]. It involves a survey of existing systems and domain experts to identify the key abstractions and mechanisms that have been employed previously and to evaluate which were successful. Behaviour analysis, on the other hand, seeks to identify objects by first understanding system behaviours, then assigning or attributing the behaviours to different parts of the system and finally recognising the parts that play a significant role in terms of their behaviours as objects. A further method for identifying objects is by use case analysis [5]. A use case is a particular pattern of usage of the system. It is a scenario that begins with some user initiating an action or sequence of events in the system. Objects in the system are identified by walking through each of the scenarios to find participating objects, their responsibilities and how they collaborate with other objects in the scenario.

1.2.3 Organising Objects

After the objects in the system have been identified, the objects need to be organised into meaningful collections by discarding unnecessary objects and mapping out relationships between the remaining objects. Objects can be grouped together into classes based on a number of criteria that stem from classification theory. Using classical categorisation, the objects that have a given property or collection of properties are grouped together into a common category or class. Clustering techniques can be used to formulate conceptual descriptions of the classes and then classifying the objects according to how closely they fit into a particular description.

1.2.4 Describing Object Interaction

To determine how the different objects fit into an object-oriented software system, different scenarios can be described showing the involvement of the object of interest and its interaction with other objects.

1.2.5 Defining Operations on Objects

An object's operations can be determined naturally when the behaviour of the object and the interface to it are considered. The behaviour of an object represents the complete set of messages that can operate on an object or to which the object can respond. A number of techniques including use cases and class-responsibility-collaborators (CRC) [6] have been successfully used during analysis not only for the discovery of objects but also to determine object operations and the interactions between objects.

1.2.6 Object-Oriented Design

The aim of object-oriented design (OOD) is to make high-level decisions about how the problem just analysed will be solved. The input to the design phase is the analysis models. Object-oriented design defines the architecture of the system and addresses implementation issues concerning object interaction, relationships between objects and choice of implementation environment and implementation language. The design process also organises the objects in the system into sub-systems and assigns these to physical processors based on the system's architecture.

1.2.7 Implementation

The implementation phase converts the design into code. An object-oriented programming language like C++, Smalltalk or Common Lisp Object System (CLOS) is preferred because they include constructs which directly realise the design into code. The objects in the analysis and design model can be directly represented in C++ or Smalltalk classes. Constructs also exist to directly represent complicated aggregation and generalisation/specialisation relationships between objects in the problem domain.

1.2.8 Object-Oriented Testing

Testing is a defined, repeatable and measurable process that is performed on the software to qualitatively demonstrate that the software functions or fails to function as specified in the requirements. Classes and objects in object-oriented programs must be tested to a greater degree than traditional programs because they are subject to reuse in projects and environments that are very different from those for which they were designed. Unit testing is carried out on individual classes to ensure that they satisfy their required behaviour. The test data can be obtained from test cases developed during the requirements determination phase. Finally, integration testing is carried out on the finished system consisting of all the different objects to demonstrate that the software meets the initial specifications.

1.3 OOD APPROACH TO ARTIFICIAL NEURAL NETWORKS

1.3.1 Identifying Objects

The main objects in an ANN can be easily identified by careful analysis of the problem statement and from domain knowledge. Careful analysis of the domain objects shows that some of the objects are aliases for previously listed objects. It is necessary to prepare a data dictionary in order to decide and eliminate unnecessary domain objects. A dictionary entry clarifies the role an object plays in the system and so reduces misinterpretations, ambiguities and name clashes. A dictionary entry for an object consists of the object name, a short description, associations with other objects and any constraints that are imposed on the object. The object's attributes and operations can also be included.

1.3.2 The Data Dictionary

- **Feedforward associative ANN**: An associative memory network where neurons have no connections to themselves or to neurons in the same layer or preceding layers. Feedforward networks may or may not have threshold elements.
- **Hopfield network**: Single layer feedback associative ANN. Neurons are not allowed to have connections to themselves even though they have connections to other neurons in the same layer.
- **Associations**: An association is a single vector or a pair of vectors to be stored in a network.
- **Exemplars**: Another name for an association, used in the context of multilayer perceptron networks.
- **Patterns**: A set of associations stored in a data file. Could be either training patterns or test patterns.
- **Weight**: The value of a connection between two neurons. A weight matrix connects layers of neurons.
- **Threshold**: A vector of values used for non-linear decision making in FANNs.
- **Training file**: The plain text or binary data file containing patterns.
- **Vector**: Data type or object created to represent a one-dimensional array of values.
- **Key**: Named vector.

Let us consider how the ANN could be developed with the OMT.

Three separate models are provided so that different aspects of the system can be visualised. The models include the object model, the dynamic model and the functional model. The object model is expressed in a class or object diagram as shown in Figure 1.6. This model shows the static structure of an ANN and serves to describe the constituent parts without the details of its operation. The class diagram shows that an ANN consists of one or more neuron layers. Each layer is modelled as a weight matrix containing a pair of activation vectors that represent the input and output activations. The activation vectors are acted on by activation functions which can be linear or non-linear in nature. The ANN uses patterns both during training and testing or validation. Patterns can thus be either training patterns or test patterns. Each pattern is just an aggregation of vector pairs which in turn consists of between zero and two vectors.

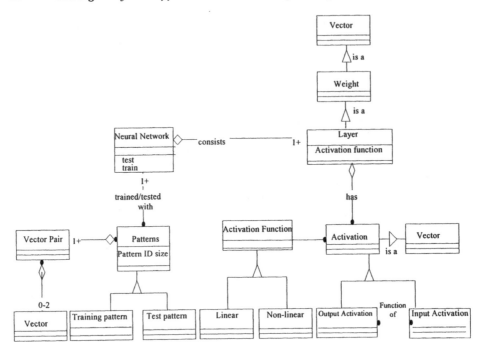

Figure 1.6 ANN class diagram

The second view, shown in Figure 1.7, represents the dynamic model of the ANN system. This view shows the different states that the system can be in, the events that the system responds to and the possible transitions between the states. The state of the system at any one time is the sum total of the values held by all its attributes and associations at that time. For a large system, the number of possible states can be very large. For modelling purposes, all the major states in the main objects are represented. While in a state, the ANN can perform certain activities until an event is received that causes a state transition. For example, in state *Training*, the ANN can *do: train, do: test* or *do: print error*.

The final view in the ANN system description represents the functional model of an ANN system. This model describes the functionality of an ANN in terms of the data it accepts and the transformations that it performs on the data. Functional models are expressed using dataflow diagrams. A dataflow diagram is a directed graph where the nodes represent functions that carry out operations and the edges represent the flow of data or resources between the functions. Dataflow diagrams exist at different levels. The highest level dataflow diagram is the system context diagram. This shows what inputs are required by a system, the sources of the inputs, the main outputs produced and, finally, the main users.

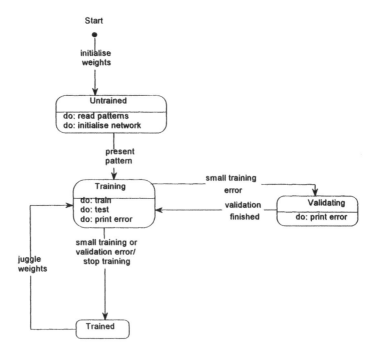

Figure 1.7 Dynamic view of a neural network architecture

The system context can be successively refined into lower level dataflow diagrams that describe the system in greater detail. At the lowest level the node transformations in the dataflow diagram degenerate into functions or algorithms that can be utilised in the finished design. The ANN context diagram is shown in Figure 1.8, while Figure 1.9 shows the dataflow diagram.

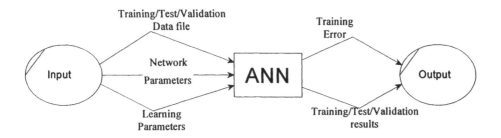

Figure 1.8 Context diagram of a typical ANN system

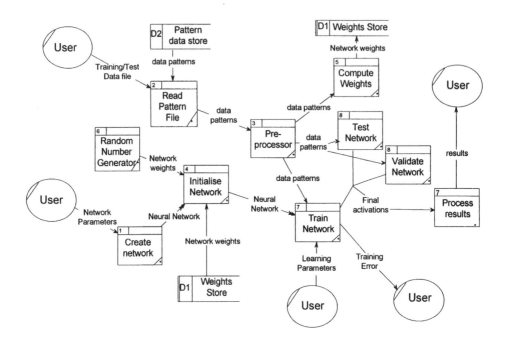

Figure 1.9 Dataflow diagram of a typical neural network system

1.4   LEARNING IN ANNS

This section discusses the key issues in ANN learning.

Given a network of neurons connected by weighted unidirectional links, some of the neurons are regarded as input neurons, others as hidden neurons and the rest are output neurons. The object of learning is to train the ANN to respond to an input vector X_p with a specified output vector Y_p. This can be accomplished by adapting the ANN weights $w = \{w_{ij}\}$ to learn the required mapping between the input and the output vectors. A qualitative mechanism describes the way in which synaptic weights are modified to reflect the process of learning [7]. Questions about what ANNs really learn have been asked since the early days of neural computing [8]. This section discusses data representations and learning in ANNs and presents the different coding mechanisms that enable ANNs to learn meaningful representations. The final part of this section presents a discussion of the different ANN learning paradigms and the associated network models and algorithms that make use of these learning paradigms.

1.4.1   Data Representations in ANNs

In general, ANN mappings are easier to establish if similar inputs give rise to similar outputs. Many application areas where ANNs have been used have involved some sort of binary representation of data. This simplifies the analysis and possibly the learning process but can result in similar inputs having quite different representations. Biological neural systems make use of a proportional form of coarse coding by way of locally tuned systems. In coarse coding systems each neuron responds to a range of input values, in between but overlapping, with those of its neighbours and the representation is said to be distributed. A given input will thus be represented by the relative activity of a number of neighbouring neuron cells. Coarse coding neurons output a 1 value if the stimulus is within their receptive field, and a 0 otherwise. Such a coding can be self-organised by the ANN. A self-organised coding has no obvious meaning to a human observer, without the use of exhaustive analysis or some forms of translation device. On the other hand, coding imposed on the ANN by an implementer can be, and usually is, meaningless to the ANN. However, if the ANN has some means of interacting with the environment of which it is a part, then the internal representations may begin to have some meaning. Data representation internally in ANNs can either be local or distributed or both. In a localised representation, inputs are mapped consistently onto a single category. In such a representation, the localised constituents of the representation can be associated with specific operations appropriate for the category. In a distributed representation, a single input causes many hidden neurons to become activated. The network can thus encode unknown feature representations with each hidden neuron responding to one or more aspects of the features in the input representation. The choice of input representation is thus important in determining the internal representations. Input representations can also be distributed, local or both. A localised representation usually produces a larger feature set but could facilitate training. A distributed representation produces more compact feature vectors but could be difficult to train, especially where similar inputs in state space tend to be represented differently. The input representation depends to a large extent on the coding strategy used. Coding strategies could either be discrete or continuous.

Discrete coding schemes assume that units can only be on or off. As such, they are relatively immune to noise. On the other hand, they are severely limited in resolution and range.

Continuous coding schemes encode the input as a triangular or Gaussian fuzzy number. Triangular fuzzy numbers are specified by their widths which determine the fraction of the number encoded by neighbouring neurons. Gaussian fuzzy numbers are specified with a given standard deviation that determines the spread of the number to the neighbouring neurons.

1.4.2 What ANNs Learn

ANNs learn representations [9]. For a particular problem, the space of possible representations is very large. Researchers into ANN learning are interested in the conditions under which a useful representation for a problem would be learned. ANN learning is based on a broad class of parametric search techniques that may be recursive, non-linear, biased and even inefficient. Still, this surpasses statistical learning which exists, in principle, only when representative samples and detailed knowledge about the environment are available [10]. The features of an ANN, including the type of neurons, the architecture and the learning rules, are all independent of the problem to be solved. This has meant that when an ANN is designed to solve a given problem, the features of the network are chosen in an *ad hoc* manner. A better understanding of what an ANN actually learns will provide a basis for selecting ANNs for solving specific problems. An integrated theory of learning and representation in ANNs is thus needed.

In ANNs, a characteristic learning task for an ANN is that of categorisation. Given a set of stimuli defined in an arbitrary feature space, an ANN can be used to sort the stimuli into pre-defined categories. Each stimulus is encoded as a unique pattern of pre-defined features. The activation of an input neuron in turn produces patterns of activity throughout the ANN. Learning consists of changes in certain properties of the hidden and output neurons. In the ANN, knowledge is stored as a pattern of connections or connection strengths amongst the neurons. The information learned directly determines how the neurons interact. Neurons have very little information stored internally. Typically, only a scalar activation level is stored. The activation level is used as a sort of short-term memory. The long-term storage of information is accomplished by altering the pattern of inter-connections or by modifying the weights associated with each neuron. There are three kinds of constraints on ANN learning:

1. The training data set is usually incomplete and erroneous. This means that the ANN must constantly update parameter estimates with data which may represent only a small sample from a possible population.
2. The conditional distribution of categories with respect to the input stimuli and features are a priori unknown and have to be determined from a sample that is unrepresentative.
3. Local information may be varying or misleading. This would lead to a poor trade off between using data available at the time or waiting for more information to become available.

Single layer perceptron networks have an inherent capability to learn any function that can be represented, but they are extremely limited in terms of what they can represent. Multilayer ANNs use hidden neurons to increase their representation capability and hence their computational power. The presence of hidden neurons allows the network to perform more complicated input-to-output mappings by constructing more complicated decision boundaries in state space. It is thus possible for an ANN in principle to be able to perform any classical computation [11].

1.4.3 Learning Algorithms in ANNs

ANNs are capable of constructing models of arbitrary systems, represented by time-varying stochastic processes over some vector spaces. This is done by learning a mapping between the input vector space and the output vector space. There are five common learning paradigms in ANNs.

1. **Auto associator:** Under this paradigm, patterns are stored by repeatedly presenting them to the ANN during learning. The network learns to represent them internally. During recall, an arbitrary pattern is presented to the ANN and the network is supposed to recover the stored pattern which is closest to the one presented. The input and output patterns span the same vector space.

2. **Pattern associator:** Under this paradigm, pairs of patterns are presented and stored in the network during learning. One of the patterns in a pair represents the key and the other represents an associated pattern. During recall, presentation of a complete or partially complete (corrupted) key should enable the network in principle to reproduce the associated pattern.

3. **Pattern classifier:** In this case, patterns are presented to the network which is supposed to categorise them according to some pre-defined set of classes. Usually, the learning process is supervised by a teacher signal. During training, several patterns are presented along with their correct classification. During recall, the network should, in theory, be able to correctly classify patterns that are different from, but similar enough to, the exemplars in each category used in the training phase. Algorithms that implement this kind of learning are referred to as supervised learning algorithms.

4. **Regularity detector:** Given an arbitrary probability distribution of patterns, each pattern is presented to the ANN with the probabilities in which they occur in that distribution. The ANN is supposed to discover the salient features in the distribution and thus classify each pattern according to these features. In this case, there is no teacher signal provided. The different categories and what they represent have to be learned. Algorithms that implement this kind of learning are referred to as self-learning or unsupervised learning algorithms.

5. **Reinforcement learning:** Under this paradigm, the network accepts input signals that are processed and transformed into output signals. The only clue to the correctness of this transformation is an extra reinforcement signal. The reinforcement signal acts to reward or penalise the ANN or learning system, depending on the success or failure of its current computation. The network adapts to try to minimise the penalties, and/or maximise the rewards. Eventually, only the correct transformations are produced.

1.4.4 Unsupervised Learning Networks

Unsupervised or competitive learning takes place in the context of sets of hierarchically layered neurons. Unsupervised learning algorithms have the property that a competition process, involving some or all of the processing elements, always takes place before each episode of learning. If the characteristics of the patterns can change slowly with time, an unsupervised learning system that can track these changes will provide better performance [10]. The available information is in the form of a set of input patterns $x^p \in X$, where $X = \{x_1, x_2, \ldots, x_n\}$ is a data set of n items. The operations that could be performed on X include:

- **Clustering:** Clustering in X means the identification of an integer c, $2 \le c < n$, and a partitioning of X by c clusters. The learning system has to find inherent clusters in the input data. The output of the system is the cluster label for an available input pattern.

- **Classification:** If S denotes the data space from which **X** was drawn, i.e. **X** $\subset$ **S**, classification in **S** is a process whereby **S** is partitioned into a number of decision regions. The unsupervised learning system delineates the decision regions in **S** in an attempt to discover the structure in **S**.
- **Vector quantisation:** In this sort of categorisation, continuous space has to be discretised. The input of the system is the p-dimensional vector, **X**. The unsupervised learning system has to find an optimal discretisation of the input space.
- **Dimensional reduction:** The input vectors have to be grouped into a subspace that has lower dimensionality than that of the data. The unsupervised learning system learns an optimal mapping that preserves the variance of the input data in the output space.
- **Feature extraction:** The unsupervised learning system is used to extract features from the input data. This usually leads to a dimensional reduction of the input space.

1.4.5 Clustering Networks

Figure 1.10 shows a simple clustering network. The network is fully connected, with weights w_{ij} connecting neuron i in the output layer to neuron j in the input layer. Both the input and weight vectors are normalised to unit length initially.

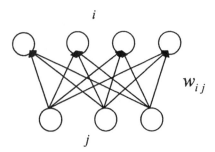

Figure 1.10   A simple clustering ANN

The activation of neuron i in the output layer is given by

$$a_i = \sum_j w_{ij}\, x_j = \mathbf{w}_i^T \mathbf{x} \tag{7}$$

    The competition process selects the neuron with the maximum activation as the winner and its activation is given a maximum value of 1. All other neurons have their activation reset to 0. The learning process adapts the weight vector of the winning neuron so that it is closer to the input vector each time an input vector is presented. Consequently, the weight vectors move towards the area that has a higher density of input vectors, thus forming a cluster. The simple clustering network has inherent stability problems as the cluster centres continue to move around indefinitely. Furthermore, a malicious input vector can completely change the centres of the clusters, causing misclassification of previous training vectors. The ART networks are an advanced implementation of the clustering network with control to provide better support for incremental learning [4].

1.4.6 Vector Quantisation

Vector quantisation algorithms are used to find natural groupings in a data set. Every feature vector is associated with a point in an n-dimensional feature space. Vectors **x** belonging to the same class are assumed to form a cluster in feature space. Vector quantisation discretises the input space so that the clusters can be separated. A vector quantisation ANN approximates a mapping of an n-dimensional input space into an m-dimensional output space. The vector quantisation algorithm presupposes that vectors belonging to the same class are distributed normally with mean μ. Feature vectors are classified on the basis of their Euclidean distance, $\|x - \mu\|$, from a pre-selected set of mean vectors. Whenever a feature vector has been wrongly classified, the mean vectors are updated by moving the correct mean towards the feature vector and the wrong mean away from it. The learning rule used to update the mean vectors is given by

$$w_{ij}(t+1) = w_{ij}(t) + \eta(o_i - w_{ij}(t)) \tag{8}$$

where η is the learning rate, taken in the direction of the input vector and o_i is the output of the network.

Vector quantisation has been combined with feedforward networks to form the counter propagation networks as shown in Figure 1.11.

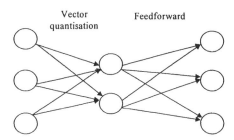

Figure 1.11 Hybrid network, vector quantisation and feedforward network

1.4.7 Supervised Learning

Supervised learning networks adapt their connection weights to learn the relationship between a set of example patterns. They are able to apply the relationship learned when presented with novel input patterns. This is because they focus on features of arbitrary input patterns that resemble those of the examples used during training. The output of the network is a function of its inputs and the connection weights between its neurons, i.e.

$$y = \Phi(\mathbf{x}, \mathbf{w}) \qquad (9)$$

where Φ is termed the discriminant function. During training, example patterns are entered as pairs of input/teacher vectors, $[\mathbf{X}, \mathbf{T}] = \{[\mathbf{x}_1, \mathbf{t}_1], [\mathbf{x}_2, \mathbf{t}_2], ... [\mathbf{x}_M, \mathbf{t}_M]\}$, where M is the number of training pairs. The network learns a mapping between the input vectors $\mathbf{x}$ and the teacher vectors $\mathbf{t}$. There are two approaches to supervised learning which differ only in the nature of the teaching signal. The first approach is based on the correctness of the solution and is termed decision-based learning. The second approach is based on the optimisation of a training or cost function and is said to be approximation based.

1.4.8 Decision-Based Supervised Learning

In this form of supervised learning the teaching signal only serves to determine whether each training pattern is correctly classified or not. Therefore binary decision vectors are used as teaching signals. The objective of training is to find a set of weights that yield a correct classification of the input patterns. For a sample classification problem, the pattern space is divided into decision regions separated by decision boundaries. The discriminant function acts as a hyperplane or hypersurface separating the decision regions. If the pattern classes are linearly separable, then a linear discriminant function can be used. Two classes of patterns are said to be linearly separable if they can be separated by a linear hyperplane decision boundary. The decision boundary is characterised by a linear discriminant function:

$$\Phi(\mathbf{x}, \mathbf{w}) = \sum_{i}^{P} w_i x_i + \theta = 0 \qquad (10)$$

The classification is decided on the basis of the values of the discriminant functions at the network's output. A binary decision is made as follows:

$$d = \begin{cases} 1 & y > 0 \\ 0 & y \le 0 \end{cases} \qquad (11)$$

The network weights are updated according to the perceptron learning rule.

$$w^{(m+1)} = w^{(m)} + \eta (t^{(m)} - d^{(m)}) x^{(m)} \qquad (12)$$

Because both $t^{(m)}$ and $d^{(m)}$ are binary, the network weights will only be updated when a pattern is misclassified in a particular epoch. If a pattern belongs to a class but is misclassified by the network, then the network weights will be reinforced by adding a fraction of the input pattern to the weights. On the other hand, if the pattern does not belong to the class but is misclassified as belonging to the class, then the weights will be anti-reinforced by subtracting a fraction of the pattern from the weights. Training stops when all patterns are correctly classified and no more weight updates take place. The learning process is said to converge. It has been proven that if the classes of patterns are linearly separable, then the learning process is guaranteed to converge to a correct solution in a finite number of steps. This is known as the Perceptron Convergence Theorem [8] and only holds when the discriminant functions are linear. Perceptron networks composed of only linear discriminant functions are severely limited because they require that the input patterns should be linearly separable if the learning algorithm converges. The perceptron learning rule has been generalised into the decision-based learning. Weight update is still by reinforced and anti-reinforced learning:

$$\Delta \mathbf{w} = \pm \eta \nabla \Phi(\mathbf{x}, \mathbf{w}) \tag{13}$$

where

$$\nabla \Phi(\mathbf{x}, \mathbf{w}) = \frac{\partial \Phi(\mathbf{x}, \mathbf{w})}{\partial \mathbf{w}}$$

is the gradient vector of function Φ with respect to $\mathbf{w}$.

1.4.9 Approximation-Based Supervised Learning

In approximation-based supervised learning the training patterns are also given in input/teaching signal pairs. The teaching signals are the desired or target values at the output nodes which correspond to a given input pattern. The objective of the learning process is to find an optimal set of weights to minimise the error between the teaching signal and the actual response of the network. The learning problem can usually be represented as a function minimisation problem. For a multilayer perceptron network with inputs vectors $x_p \in \Re^N$ and target vectors $t_p \in \Re^k$ the error due to the p th pattern vector is given by

$$E_p = \sum_k (t_{pk} - o_{pk})^2 \tag{14}$$

where

$o_{pk} = f(net_j) = f(\sum_i w_{ji} a_i + \theta_j)$ is the actual output of the k th neuron,

t_{pk} is the neuron's desired or target value,

$f(net_j)$ is the neuron activation function,

net_j is the net input to the j th neuron,

a_i is the activation of the i th neuron and

θ_j is a learnable bias weight.

E_p can be expressed in vector notation as

$$E_p = (\mathbf{t} - \mathbf{o})^{\mathrm{T}} (\mathbf{t} - \mathbf{o})$$
$$= (\mathbf{t} - \mathbf{o})^{\mathrm{T}} \mathbf{I} (\mathbf{t} - \mathbf{o}) \tag{15}$$

where I is the identity matrix. E_p is a quadratic error function and minimisation of $E_p = E(\mathbf{w})$ will provide an optimal set of weights for a given problem [12]. Gradient descent methods or methods of steepest descent can be used to solve the learning problem. These methods transform the minimisation problem into an associated system of first-order ordinary differential or difference equations.

$$\frac{dw_j}{dt} = -\sum_{k=1}^{K} \mathbf{u}_{jk} \frac{\partial E_p}{\partial w_j} \tag{16}$$

with initial conditions $w_j(0) = w_j{}^{(0)}$. This equation can be written in compact vector form as follows:

$$\frac{d\mathbf{w}}{dt} = -\mathbf{u}(\mathbf{w}, t) \nabla_{\mathbf{w}} E_p(\mathbf{w}) \tag{17}$$

where $\mathbf{u}(\mathbf{w}, t)$ is a symmetric positive definite matrix called the learning matrix. In the steepest descent method, $\mathbf{u}(\mathbf{w}, t)$ is assumed to be a unitary matrix multiplied by a positive constant η, which is the learning rate. The above system of differential equations can be shown to be stable by considering the change of the error (energy) function E_p with time:

$$\frac{dE_p}{dt} = \sum_{j}^{N} \frac{\partial E_p}{\partial w_j} \frac{dw_j}{dt} \tag{18}$$

By enforcing the condition that $\mathbf{u}(\mathbf{w}, t)$ is symmetric and positive definite, the following condition holds:

$$\frac{dE_p}{dt} = -[\nabla_w E_p(\mathbf{w})]^{\mathrm{T}} \mathbf{u}(\mathbf{w}, t) \nabla_w E_p(\mathbf{w}) \leq 0 \tag{19}$$

The above condition guarantees that the error function decreases in time and eventually converges to a stable local minimum. The speed of convergence to the minimum depends on the choice of values for the learning matrix $\mathbf{u}(\mathbf{w}, t)$.

The discrete-time version of the gradient descent method is

$$\mathbf{w}^{(k+1)} = \mathbf{w}^{(k)} - \eta^{(k)} \nabla_{\mathbf{w}} E_p(\mathbf{w}^{(k)}) \qquad (20)$$

where $\eta^{(k)}$ is the learning step size that should be bounded in a small range to ensure the stability of the algorithm. In most implementations of the steepest descent learning algorithm, $\eta^{(k)}$ is determined at each time step by one-dimensional line search as the value of $\eta \geq 0$ that minimises

$$\psi(t) = E_p(\mathbf{w}^{(k)} - \eta \nabla_{\mathbf{w}} E_p(\mathbf{w}^{(k)})) \qquad (21)$$

1.4.10 Backpropagation Learning Algorithm for Multilayer Perceptron Networks

Multilayer perceptron networks consists of layers of neurons connected by synaptic weights. An example of a fully connected network is shown in Figure 1.12.

The dynamic gradient descent algorithm described above can be used to minimise the sum-of-squares errors. This method requires the computation of the gradient $\nabla_{\mathbf{w}} E(\mathbf{w})$ for all the weights in the network during each learning iteration. This is very inefficient and computationally very expensive. It is also impracticable for large networks. The backpropagation algorithm offers an effective approach to the computation of gradients and hence a relatively efficient speedup to the training of multilayer ANNs. The backpropagation algorithm makes use of a recursive formulation to compute the error at the output of lower layer neurons when the error at the output neurons is known. Hence the computation of gradients of the energy function with respect to the lower layer weights is thus avoided. The network weights are then updated according to the Generalised Delta Learning Rule [13].

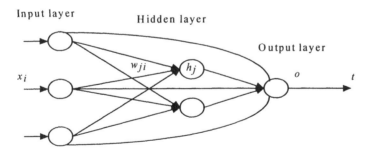

Figure 1.12 A two-layer fully connected multilayer perceptron network. There are three neuron layers and two weight layers. The number of layers in the network usually refers to the number of weight layers

1.4.10.1 Supervised Learning Parameters

Certain formulations have to be adopted when supervised learning networks are applied to solve real problems. The selection of the learning parameters has a significant effect on the network performance. The learning parameters are varied and very problem dependent. Some of the factors that influence learning in supervised networks include:

- **Learning rate η.** This specifies the step size that is taken in the downhill slope or gradient in weight space, when weights are updated. The learning procedure requires that the weight change be proportional to $\nabla_w E(w)$. True gradient descent requires that infinitesimal steps are taken. The learning rate acts as the constant of proportionality. In practical applications the largest learning rate that does not lead to oscillations during training should be chosen. Alternatively, learning rate adaptation or optimal learning methods can be used to determine the optimal learning rate at each time step.
- **Momentum term:** To prevent oscillations in the learning process, the weight update equations are usually modified to include a momentum term. The momentum term specifies the fraction of the previous weight change that is added to the weights during update. The momentum term helps to keep the weight changes going in the same direction and prevents oscillations when reasonably high learning rates are used. This enables the network to converge to a final solution much faster.
- **Range of initial weights:** Gradient descent methods require the provision of initial conditions to solve the system of dynamic equations that represent the ANN. The conditions are provided by initialising the network weights to small random values. This represents the starting point of the iterative descent algorithm. Large values for the initial weights are not recommended because the ANN could be initialised to a state that is either unstable or far away from the final solution state. This will lead to very long training times and possible difficulties in learning the required mapping. Techniques have been proposed to statistically control weight initialisation based on the training patterns [14].
- **Frequency of weight update:** An iteration involves a single training pattern presented to the system. An epoch or sweep covers the presentation of an entire block of training data to the system. The network weights can be updated after each training pattern is presented, i.e. after each iteration. This is known as data-adaptive updating and provides very fast learning and better response for on-line or real-time learning applications. On the other hand, it is numerically unstable for large problems and extremely sensitive to network learning parameters and noise effect on individual patterns. Block-adaptive methods, on the other hand, only update the network weights after one epoch, i.e. the presentation of a block of patterns or all the training patterns to the network. The network tends to be slower learning, but learning is more predictable and robust as the training step is averaged over all the patterns.
- **Validation during training:** There are methods to help monitor the training procedure. One method is to validate the training after a certain number of iterations. This is called validation. The total data for training and testing can be

composed to produce three sets of patterns: training, testing and validating. If the amount of data is not enough for training due to the complexity of the problem, then cross-validation can be used to make the best use of data. This technique splits the total data set into S subsets each of equal size. For each subset, that subset is chosen to be the test set and the other S-1 subsets are combined to form the training and validation sets.

- **Scaling and normalisation of input patterns:** Normalisation or scaling is necessary for gradient descent algorithms that have no self-normalising qualities. Normalisation or scaling of the input patterns can significantly speed up numerical convergence of the gradient descent algorithm [12]. The simplest form of normalisation is to convert all input vectors to unit length. This tends to destroy the variance of the different features that make up the input vector. The input vectors can also be scaled so that the variance between the features in each input vector is preserved [15].

- **Network architecture:** The generalisation capability of a multilayer network depends on the architectural parameters. This includes the number of weight layers, the number of hidden neurons, the type of the discriminant functions and the nature of the activation functions. The ability of the network to generalise on the training set is very sensitive to the number of hidden neurons. A rule of thumb for obtaining good generalisation in backpropagation networks is to use the smallest network that will fit the data. If the number of hidden neurons is too large, the network will take too long to train and will memorise instead of generalising on the training data. The test performance on noisy input or patterns not seen before will thus be very poor. On the other hand, if the number of hidden neurons is too small, the network will not be able to learn or will be very sensitive to the network initial conditions. Small networks are also more likely to become trapped in a local minimum. A number of techniques have been proposed for determining an optimal number of hidden units to learn a particular task. These include algebraic projection analysis [16], network growing and pruning [17] and network evolution [18]. The nature of the discriminant functions coupled with the kind of activation function mostly affects quality of the classification performed by the ANN. Conventional multilayer networks use linear discriminant or linear basis functions with linear or non-linear activation functions. As mentioned above, such networks produce hyperplane decision boundaries to separate classes of input patterns. More recently, multilayer perceptron networks using radial, elliptic, spline or wavelet basis functions with Gaussian activations have been proposed [19]. Radial basis function networks have a great advantage in that they are much faster to train than multilayer perceptrons using backpropagation and are thought to be more suitable for approximating and learning smooth continuous functions from sparse data. The main reason is due to the fact that while multilayer perceptrons construct hyperplane decision boundaries in order to separate class categories, the corresponding decision boundaries are hyperspherical or hyperellipsoidal when radial basis networks are used. The resulting classification is much more accurate and can be constructed much faster because of typical clusters found in data.

1.4.11 Speeding up Supervised Learning

The greatest single obstacle to the application of multilayer networks trained by backpropagation in real-world problems is the slow speed at which current algorithms learn. Even on relatively simple problems, standard backpropagation often requires a lengthy training process in which the complete set of training exemplars is presented a large number of times. One solution is to run the network on even faster computers, but the sequential nature of conventional digital computers is a limiting factor to the speed at which the networks can be run. Hardware implementation using VLSI and optics [20] with programmable interconnections would significantly increase the speed of computation. Faster learning variations of the backpropagation algorithm such as quickprop and rprop [21] have been proposed to overcome some of the speed limitations. Second-order methods such as conjugate gradients and generalised projections [22] can search for the minimum faster than conventional gradient descent used by backpropagation.

1.5   ANALYSIS, DESIGN AND IMPLEMENTATION OF ANNS

The current generation of ANN systems exist mostly as legacy software or code libraries that have been programmed to solve specific sets of problems. Despite the prevalence of general purpose ANN software both in the public domain and commercially, there has been little or no attempt to create a robust software architecture for ANN systems which can be integrated into general commercial and industrial systems development. Some notable efforts include where dataflow diagrams were presented for ANN operations. This section describes how a robust object-oriented architecture for ANNs can be constructed. Such an architecture will allow ANNs and other intelligent systems to be widely used as components in general purpose commercial and industrial systems development efforts, whether they are in power system monitoring and protection, power system control, industrial plant control systems, financial software systems or datamining and visualisation systems in distributed databases including the World Wide Web. With a clearly defined and open architecture, it will be possible for system designers to directly incorporate ANN components in their designs or call on other intelligent systems as part of a distributed object system using an object middleware mechanism such as Microsoft's object linking and embedding (OLE). The implications of such an architecture on the future use of ANNs are phenomenal. Self-learning or trained ANNs that aid intelligent decision making will become a standard part of many new software packages as commercial software designers seek to incorporate intelligence into their products. ANNs will become an indispensable utility as more people seek software with a minimum level of intelligence that can adapt to their everyday needs. More innovative and practical uses will be developed for ANNs as more people get to understand the advantages and limitations of the technology.

1.5.1 Software Architecture

All software systems should have an architecture so that they can be easily integrated with existing systems or easily modified or customised when the requirements of an organisation or application change. In [23], architecture is described as the structuring paradigms, styles and patterns that describe a software system. A software architecture proclaims and enforces system-wide rules regarding the organisation and access of data and the overall control of the software system. The software architecture serves a number of extremely useful purposes. Some of the more common ones are as follows:

- It acts as a basis for communication between different designers and between designers and users of a software system.
- It serves as a high-level documentation of the system and as a starting point from which the effectiveness of the system can be measured or changed.
- It serves to provide users of the system with a certain degree of confidence about the robustness of the system and ensures that their investment in the software is protected.
- It acts as a boundary that delimits both the expectations and the implementation of the software system.
- It serves as a starting point from which modifications or maintenance changes to the system are made.

In this chapter, a new approach to constructing ANN systems based on object-oriented techniques is presented. Detailed descriptions of the analysis and design process are presented. The approach makes use of the object modelling technique as a basis for the development of software ANN architectures. An object-oriented architecture typically derives its properties from the classes and class hierarchies found within class clusters in the problem domain.

1.5.2 System Development

Software systems development is a very complicated task. The complexity of software is due to the fact that, usually, the developers of a software system are different from eventual users of the system and also from the people who have to maintain the system. The system has to be developed from a problem statement where the true requirements are often intermixed with design decisions [24]. Most of the time the problem statements are ambiguous, inconsistent or even wrong because the users are unsure of what the final system will do. Some requirements, although precisely stated, will have unpleasant consequences on the system behaviour or will impose unreasonable implementation costs. The complexity is further increased by the very flexible nature of the software itself, which can be constructed in an almost unlimited number of ways. This leads to high maintenance costs because any changes to the system will require a thorough understanding of the structure of the system if unpleasant side-effects are to be avoided. The complexity has to be reduced or handled in an organised way if reliable, robust and maintainable software systems are to be constructed [25].

This is done by a process of decomposition where both the software system and the development process are divided into smaller, more manageable parts. The software development process can be divided into a number of phases as shown in Figure 1.13. The product of each development phase is a model or view of the system in miniature represented by a document or set of documents that become the input to the next development phase [26].

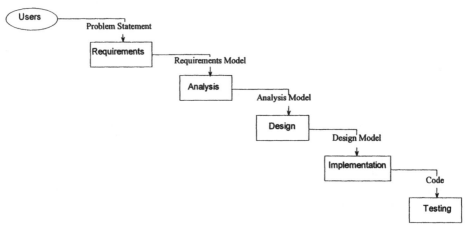

Figure 1.13   Phases in software systems development

1.5.3   Software Life-Cycle Models

The term software life-cycle can be defined as a model used to help explain and understand the software development and maintenance process. A software life-cycle model describes how the software development process progresses through the different phases. The aim of life-cycle models is to reduce the inherent complexity in the software development process in an attempt to deliver correct, reliable and robust software systems on schedule and according to budget. Different life-cycle models have been proposed with different characteristics. Three of the most common of these life-cycle models are briefly described below.

1.5.4   The Waterfall Life-Cycle Model

A schematic diagram of the waterfall model is shown in Figure 1.14. In the waterfall model, each phase has to be completed before the next phase can begin. Sometimes, the phases are allowed to overlap but the ordering of phases is strictly maintained.

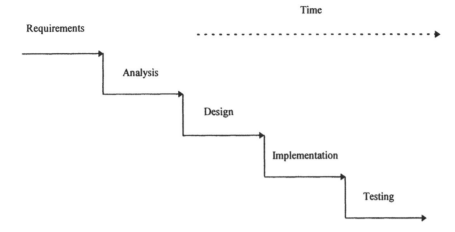

Figure 1.14   The waterfall life-cycle model of software development

1.5.5 The Evolutionary Life-Cycle Model

The development of a large software system is a slow process that can take a long time to finish. Where the requirements are completely known at the start of the project, the waterfall model can be easily applied to develop the system. In most cases the requirements cannot be completely determined at the start of the development process. Such systems are better developed in a step-by-step manner beginning with a few of its core functions. As the understanding of the system functions evolve, new functions can be added. In this way, the system is incrementally enlarged until the desired level of functionality is attained. Such an incremental strategy also provides faster feedback during the development process. In practice, the system can be divided into parts according to requested services. The completion of each part extends the system functionality up to the finished product which comprises the whole of the system functionality. Figure 1.15 shows an evolutionary life-cycle model.

1.5.6 Prototyping

Where it is difficult to determine how a system is supposed to work either due to technical or functional reasons it is helpful to develop a prototype of the intended system. This is especially true when user interfaces and highly interactive systems are to be developed. The prototype focuses on the properties of the system that require further insight. It allows the system developers to experiment with a number of design options and thus serves as a complement to incremental system development [25]. Prototyping is a useful technique for understanding an application and can act as a means of communication between the developer and the eventual users of a software system. Prototyping differs from incremental development in that the aim is not to create a fully working product, but to emphasise and demonstrate certain aspects of the intended system.

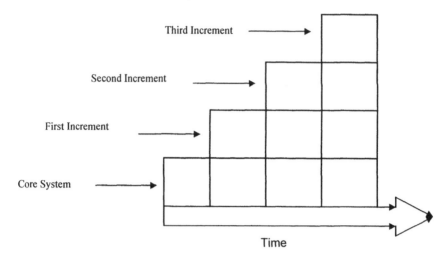

Time

Figure 1.15   Incremental software delivery using the evolutionary life-cycle model

1.5.7   Phases of Software Development

The different phases in the software systems development process are briefly described below.

- **Requirements definition:** Requirements definition or requirements specification is usually required at the start of any software development process. A requirements document can be developed from information about the environment in which the software system will be used. A need for a new system or modification of an existing system is identified by users of the system. This leads to the creation of a problem statement which becomes the starting point for the software development process.
- **Analysis:** Analysis is the study of a problem prior to designing a solution for it. The aim is to build a problem model, i.e. to create a description of what is required and what will eventually be built without attempting to say how it will be built. The product of the analysis process is a document or sets of documents that describe the problem in a clear, unambiguous and easily understood manner. The analysis document represents the analysis model. These analysis documents are completely in the problem domain and the vocabulary used in the description is consistent with those found in the problem statement. This allows the eventual users to easily understand and comment on the analysis model so that the correct system can be designed and subsequently implemented.
- **Design:** The input to the design phase is the analysis model. The aim of the design is to construct a solution model by deriving a high-level strategy for solving the problem and building the solution. The process of design will refine the analysis model to take advantage of a particular implementation environment, which consists of the available hardware, operating system, network resources, programming languages and database management systems

[1, 27]. The design phase is entirely in the solution domain and serves to structure and organise the software system to give it an overall architecture. Two main stages of design can be identified: a systems design stage and a detailed design stage. In the systems design stage, high-level strategic decisions are taken about the overall architecture of the system, the implementation environment and the mixture of hardware and software required to realise the system. The different subsystems that make up the complete system are also identified and organised. In the detailed design stage, lower level decisions that directly affect the implementation of the systems such as choice of data structures and efficiency of algorithms and algorithm designs are carried out.

- **Implementation:** The goal of the implementation phase is to realise the system according to the blueprints set out in the design phase. This can be done using a programming language or database management system.
- **Testing:** The input to the testing phase is the finished program code or packages from the implementation phase. In the testing phase, the system is checked to make sure that the performance of the system meets the specifications laid out in the requirements document.

1.6 OBJECT-ORIENTED ANALYSIS AND DESIGN OF FEED FORWARD ANNS (FANNS)

This section presents the application of object-oriented techniques to the analysis and design of FANNs. The object modelling technique (OMT) is applied to the creation of both the analysis and design models. The finished design is implemented using the C++ programming language. Unit testing is carried out on the individual classes to ensure correct behaviour of the ensuing objects. Finally, integration testing is carried out on the finished ANN using both random data and standard test data.

1.6.1 FANNs

FANNs are a class of fixed weight ANNs. These networks have the common property that the weights and/or thresholds (bias) are usually pre-calculated and pre-stored. The inputs to the networks can be binary values (0/1), bipolar (-1/1) or real numbers. The weights are derived from the input patterns usually by taking a correlation of the pairs of input vectors. Processing of the input vectors to compute the weight is similar, for both bipolar and real values. A slight modification is required for binary values. The computed weights are pre-stored in a weight collection of vectors (matrix) for use during testing.

A FANN can be used to store and retrieve associations. An association is a vector or a pair of vectors that are presented at the input to the network. Associations can be binary, bipolar or real valued. The network operates in three phases: a storing phase, a validation phase and a retrieving phase. During the storing phase, the network is presented with a training file containing a set of associations to be stored. These are the patterns or exemplars that the network is required to learn. In order to monitor proper storing, a validation phase is introduced. During validation a number of unseen data to the FANN are tested to confirm proper training. During the retrieving phase one part of an association (the key) is presented and the network has to retrieve the second part of the association. The key could simply be a corrupted version of one of the stored vectors, in which case a holographic retrieval can take place.

1.6.2 Organising the Objects

Objects do not exist in isolation. The next step in the analysis process is to organise the domain objects and map out the associations between them. The domain objects and their relationships can be represented by a domain object model. A domain object model describes the static structure of objects and the relationships between objects found in the problem domain. Furthermore, objects can be added which do not directly relate to the problem to be solved but are required to provide operations and services to the domain objects. These are called system objects and the combination of system and domain objects forms the system object model. The FANN is described as consisting of one or more weight layers. In the dictionary description, a weight is considered to be a single value. Domain knowledge shows that the weights between layers of neurons can be represented by a matrix or a collection of vectors. Even though matrices are not domain objects, they are still valid objects in the system object model as they help to efficiently describe and express the structure of the system. A similar explanation can be used to justify a generalisation relationship between vector and threshold. For each neuron, there can be only one threshold value. For a layer of neurons, all the threshold values can be collected to form a vector. The vector representation is more efficient in terms of expression when communicating design decisions. In the design, an associative ANN object has been created which forms the basis for all other types of associative memory networks. This object defines the interface and representation of FANNs. The interface and representations are inherited and then refined by adding the extra operations and data required to exhibit a specific behaviour.

1.6.3 Describing Object Interactions

Object-oriented systems are made up of collections of autonomous or semi-autonomous objects that interact or co-operate to solve a specific problem. Object interactions include the set of messages that an object should respond to and the set of behaviours that an object can exhibit. Object interactions can be discovered using scenarios or use cases. A scenario is a sequence of events that occur during one particular execution of a system. Scenarios use event trace diagrams to describe sequences of events and the objects exchanging those events during a particular execution of the system. For example, a *train* scenario can show the sequence of events that take place when the network is used to store a set of patterns.

1.6.4 Defining Operations on Objects

Event trace and object interaction diagrams aid in refining the relationships between different objects. Most of the operations required can also be obtained
from an examination of the flow of events to and from the object. This, however,
does not represent the complete picture. Where an object has very complicated
dynamic behaviour, all its operations cannot be captured simply by using event
trace diagrams. In such cases, a state-transition diagram can be used to represent
a dynamic model of the system. A state diagram for an object is a graph whose
nodes represent the different states in which the object can be in, and whose directed edges are transitions between the states. For the FANN, a dynamic model
has been developed and shown in Figure 1.16.

The dynamic model shows the need for an *initialise* member function which
uses requested network parameters to initialise the network. Extra functionality
and further objects are also required to obtain and process network parameters
from the user and also for pre-processing of data patterns and post-processing of
test results if necessary. Finally, a functional model can be used to describe what
happens in the ANN. The functional model shows what the different inputs to an
FANN are and the computation required to transform these inputs into output
values. The functional model and dataflow diagram are shown in Figures 1.17
and 1.18, respectively.

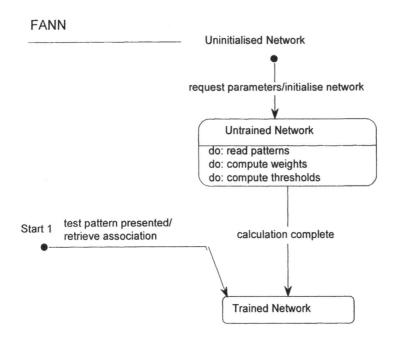

Figure 1.16 Dynamic model of a FANN

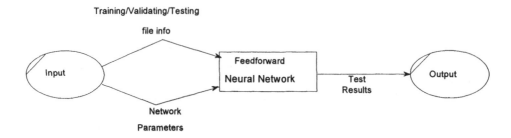

Figure 1.17   Functional model of a FANN

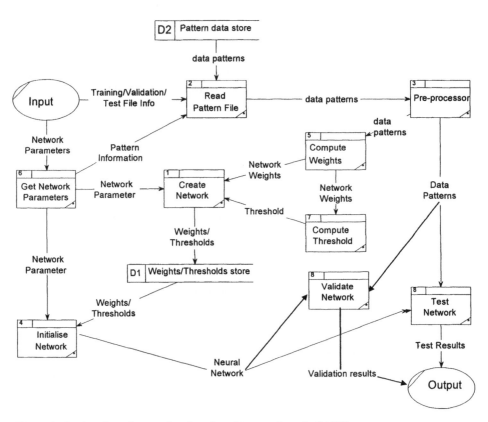

Figure 1.18   Dataflow diagram that describes the operation of a FANN

1.6.5   ANN Systems Design

In the design process, an architecture for the FANN is produced on the basis of information contained in the different models constructed during analysis. The design process consists of partitioning the ANN system into subsystems that perform specific functions in the context of the ANN. In this design, the following subsystems have been identified:

- The input subsystem.
- The data pre-processing subsystem.
- The training/testing subsystem.
- The output subsystem.

The input subsystem consists of object(s) that handle interactions with the user to obtain network parameters and information relating to the data patterns. The data pre-processing subsystem consists of objects that handle the reading of desired training/testing patterns from a disk file or other media such as sensors and its subsequent pre-processing. The training/testing subsystem includes objects that make up the ANN, weight initialisation, reading and writing of network weights and final activation values. The output subsystem includes objects that handle the post-processing of activation values and test results to produce statistical information about the performance of the ANN. The flow of information between the different subsystems in the ANN is shown in Figure 1.19. This architectural description is applicable to the design of most classes of ANNs. The different subsystems are layered on top of the operating system to create the complete ANN system, as shown in Figure 1.20. Such layering of partitions constitutes a horizontal decomposition of the ANN system. It is also possible to decompose the ANN system vertically into weakly coupled subsystems. The choice of partitioning depends on the implementation environment which includes the operating system, hardware, software and network resources.

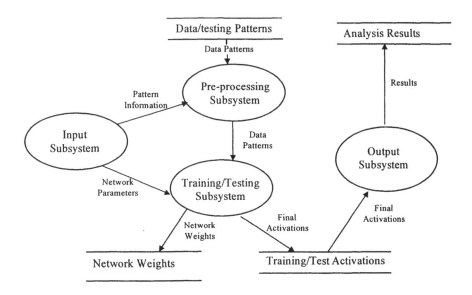

Figure 1.19 Information flow between the different subsystems in an ANN

1.6.6 Implementation and Testing

The finished design has been realised using the C++ programming language. The different objects that make up the system have been implemented using the class mechanism provided in C++. Because the ANN has been developed as an object-oriented program, testing takes a different approach from that used in conventional systems design. Each object exposes a certain behaviour, which is expressed in the C++ class. The classes are divided into objects which can then be independently tested for the required behaviour. Tested objects are then pooled together into the different subsystems for testing. Finally, the subsystems are pooled together into the complete ANN system for an integration test. This form of testing relies on the fact that the objects that make up the system are independent entities which can exist outside the system. Thus their creation and testing are completely decoupled from the system development process. This leads to more robust architectures and eliminates the need for exhaustive testing on a grand scale to prove correctness.

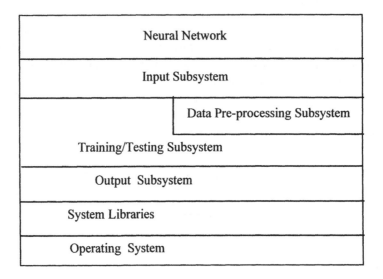

Figure 1.20   System block diagram for an ANN

1.7 CONCLUSIONS

This chapter has demonstrated a scheme for presenting ANNs using conventional diagramming techniques. The method is based on a standard notation with a few easily recognised symbols as described by the object modelling technique (OMT). For any given problem, if a suitable representation can be found, then it is possible to configure an ANN to learn the representation and hence solve the problem. This implies that an ANN solution is possible for any problem where a suitable representation can be found. The implications for ANN designers are potentially very severe as difficult learning problems can be construed as poor solution design rather than any inability on the part of the ANN to solve the problem.

In this chapter, neural network software development has been approached from a system development perspective. The analysis methods used above can be applied to the development of most neural network architectures. The software architecture makes the completed system easier to understand and extend. It also provides a framework in which modifications in the design or implementation can be incorporated so that the system does not disintegrate when necessary maintenance or upgrade is carried out. The software architecture makes it easier to incorporate neural network technology into other industrial systems development efforts by reducing the effort required from non-technical and non-expert application designers wishing to incorporate neural technology into their designs.

1.8 REFERENCES

[1] J Rumbaugh, M Blaha, W Premerlani and W Lorensen, *Object-Oriented Modelling and Design*, Prentice-Hall, 1991.

[2] A Cichocki and R Ubenhauen, *Neural Networks for Optimisation and Signal Processing*, John Wiley & Sons, 1993.

[3] J Moody and C Darken, 'Faster learning in neural networks of locally-tuned processing units', *Neural Computation*, 1, 1989, 281-294.

[4] J Freeman and D Skapura, *Neural Networks: Algorithms, Applications, and Programming Techniques*, Addison-Wesley Publishing Company, 1991.

[5] R Bindu, *Object-Oriented Databases: Technology, Applications and Products*, McGraw-Hill, 1994, pp.207-224.

[6] N Wilkinson, *Using CRC Cards: An Informal Approach to Object-Oriented Software Development*, SIGS Books, 1995.

[7] D Hebb, *Organisation of Behaviour*, John Wiley & Sons, 1949.

[8] M Minsky and S Papert, *Perceptrons*, MIT Press, 1988.

[9] J Hanson and D Burr, 'What connectionist models learn: learning and representation in connectionist neural networks', *Brain and Behavioural Sciences*, 13, 1990, 471-518.

[10] R Duda and P Hart, *Pattern Classification and Scene Analysis*, John Wiley & Sons, New York, 1973.

[11] K Hornik, M Sticchcombe and H White, 'Multilayer feedforward networks are universal approximators', *Neural Networks*, 2(5), 1989, 259-366.

[12] D G Luenberger, *Linear and Nonlinear Programming*, Addison Wesley, 1984.

[13] D Rumelhart, G Hinton and R Williams, 'Learning internal representations by error back-propagation', in D Rumelhart and J McClelland (Editors), *Parallel Distributed Processing*, MIT Press, 1, Chapter 5, 1986, 152-193.

[14] G Drago and S Ridella, 'Statistically controlled weight initialization (SCAWI)', *IEEE Transactions on Neural Networks*, 3(4), July 1992, 983-986.

[15] L L Lai, F Ndeh-Che, Tejedo Chari, P Rajroop, and H S Chandrasekharraiah, 'HVDC systems fault diagnosis with neural networks', *Proceedings of the 5th European Conference on Power Electronics and Applications*, The European Power Electronics Association, 8, Sept 1993, 145-150.

[16] S Kung and J Hwang, 'An algebraic projection analysis for optimal hidden units, size and learning rates in backpropagation learning', *IEEE International Conference on Neural Networks*, ICNN '88, San Diego, **1**, 1988, 363-370.

[17] R Reed, 'Pruning algorithms, a survey', *IEEE Transactions on Neural Networks*, **4**, 1993, 740-747.

[18] C Jacob and J Rehder, 'Evolution of neural net architectures by a hierarchical grammar-based genetic system', in R F Albrecht, C R Reeves and N C Steele (Editors*), Artificial Neural Nets and Genetic Algorithms*, Springer-Verlag, August 1993, 72-79.

[19] J Park and I Sandberg, 'Approximation and Radial Basis Functions', *Neural Compuation*, MIT Press, 1993, 305-316.

[20] E Georges, L L Lai, F Ndeh-Che and H Braun, 'Implementation of neural networks with VLSI', *Proceedings of the Fourth International Conference on Neural Networks*, IEE, UK, June 1995, 489-494.

[21] J Nanda, A Sachan, L Pradhan, M L Kothari, A K Rao, L L Lai and S M Prasad, 'Application of artificial neural network to economic load dispatch', *Proceedings of the Fourth International Conference on Advances in Power System Control, Operation & Management*, IEE, Pub No 450, 1997, 707-711.

[22] S Yeh and H Stark, 'A fast learning algorithm for multilayer neural networks based on projection methods', in J Mammone (Editor), *Neural Network, Theory and Applications*, Academic Press, 1994, 323-342.

[23] S Gossain, 'The emergence of the system architect', *Object Expert*, **1**, SIGS Publications, 1995, 58-60.

[24] L Peters, *Advance Structured Analysis and Design*, Prentice-Hall International Editions, 1988.

[25] I Jacobson, M Christerson, P Johnson & G Övergard, *Object-Oriented Software Engineering: A Use Case Driven Approach*, Addison-Wesley, 1992.

[26] E Yourdon, *Modern Structured Analysis*, Prentice-Hall International Editions, 1989.

[27] J Coplien, *Advanced C++ Programming Styles and Idioms*, Addison-Wesley Publishing Company, 1992.

2

Evolutionary Computation

In this chapter, evolutionary computation (EC) is explained. Techniques used in genetic algorithms and evolutionary programming are introduced. These two evolutionary algorithms are then implemented by following the object-oriented analysis, design and testing procedures.

2.1 INTRODUCTION

This chapter presents a general overview of evolutionary algorithms (EAs), in particular evolutionary programming (EP) and genetic algorithms (GAs). Evolutionary algorithms (EAs) are computer-based problem-solving systems based on principles of evolution theory. Different EAs have been developed and they all share a common conceptual base of simulating the evolution of individual structures via the processes of selection, mutation and recombination. The processes depend on the perceived performance of the individual structures as defined by an environment. The interest in these algorithms has been rising fast because they provide robust and powerful adaptive search mechanisms. The interesting biological concepts on which EAs are based also contribute to their attractiveness. This way, over the last few years we have been witnessing a considerable increase in research based on EAs that has been documented in a large number of conference proceedings, journal articles, research reports and working papers. However, much of these research is not widely available in comparison to amount of work on artificial neural networks (ANNs). Believing that EAs have an immense potential for applications in the field of power systems, this chapter is intended to provide an overview of the research being done in this area; furthermore, this chapter also includes a classification of available EP methods and discusses their issues. The field of evolutionary computation is referred to by some as computational intelligence (CI) and by others as soft computing.

The most popular EAs developed so far are the following:
- genetic algorithms;
- evolutionary programming;
- evolution strategies;

- genetic programming;
- classifier systems; and several other problem solving strategies such as simulated annealing.

These EAs are useful for optimising problem solutions when other techniques like gradient descent or direct analytical discovery are not possible. EAs based upon biological observations date back to Charles Darwin's discoveries in the nineteenth century: the means of natural selection and the survival of the fittest, i.e. the 'theory of evolution'. The resulting algorithms are thus termed EAs.

GAs are the most popular and widely used of all the evolutionary algorithms. However, in recent years EPs seem to have caught up with GAs. GAs have been widely applied to solve complex non-linear optimisation problems in a number of engineering disciplines. Like other computational intelligent systems such as neural networks and fuzzy systems, GAs currently exist as lines of computer code written to solve particular problems. In this chapter, a similar analysis and design philosophy, using the object-oriented approach, is presented for the construction of GAs and EPs. The use of object-oriented techniques means that an EP component can be created and incorporated into other designs where an evolutionary solution is required. The high level of reuse exemplified by such an approach is demonstrated in the following chapter by the construction of an EP-ANN/GA - ANN system incorporating the EP/GA design.

2.2   GENETIC ALGORITHMS (GAS)

Genetic algorithms (GAs) are robust search mechanisms based on the principle of population genetics, natural selection and evolution [1]. GAs perform a global search on the solution space of a given problem domain [2]. Genetic search is capable of satisfying strong hidden constraints with reasonable efficiency. The characteristics of genetic search mechanisms enable them to search globally and still converge to a solution. The characteristics include:

- **Learning while searching:** As the GA search progresses, the scope of the search is narrowed as information about the function space accumulates.
- **Sustained exploration:** The GA incorporates mechanisms that prevent the search space from being irrevocably narrowed so far as to let the optimal solution states slip permanently through the GA search net.

Genetic search is characterised by three components:

1. **The ongoing state:** The ongoing state represents knowledge that contains information previously acquired during the search and also a means of feeding information from past searches to aid the current or future search.
2. **The search function:** The search function uses the ongoing state at time (generation) k to generate the next point to search.

3. **The learning Function:** The learning function uses the ongoing state and the location and value of the most recently searched point to update the ongoing state.

Search strategies can broadly be divided into three categories [3]:

- **Path-based models:** In path-based models the search space is recursively defined by a starting state and a set of state-to-state transition operators. The search strategy then has to find a solution state and a path to it from the starting state. Examples of path-based search include tree searching techniques such as depth first search.
- **Point-based models:** Point-based models maintain the location and value of one point in the search space defined to be the current point. A neighbourhood is defined around this point to be the promising region. The current point is used as the standard for comparison. New points are generated at each step of the search. Better points get credit by becoming the current point while bad points are discarded.
- **Population-based models:** Population-based models maintain a set of locations and values in the function space. The average value of the population can be used as the standard of comparison and a discovered good point can be added to the population according to some rule or rules.

2.2.1 Features of GAs

Genetic algorithms belong to the class of population-based search strategies. They operate on a population of strings (chromosomes) that encode the parameter set of the problem to be solved over some finite alphabets. Each encoding represents an individual in the GA population. The population is initialised to random individuals (random chromosomes) at the start of the GA run. The GA searches the space of possible chromosomes (Hamming space) for better individuals. The search is guided by fitness values returned by the environment. This gives a measure of how well adapted each individual is in terms of solving the problem and hence determines its probability of appearing in future generations. A binary encoding of the parameters of the problem is normally used. It has been mathematically proven [1] that the cardinality of the binary alphabet maximises the number of similarity templates (schemata) on which the GA operates and hence improves the search mechanism.

Two types of rules are used by GAs in their search for highly fit individuals: selection rules and combination rules. The selection rule is used to determine the individuals that will be represented in the next generation of the GA. The combination rules operate on selected individuals to produce new individuals that appear in the next generation. The selection mechanism is based on a fitness measure or objective function value, defined on each individual (chromosome) in the population. Two major selection mechanisms are commonly adopted in the GA search: roulette wheel selection and tournament selection. In roulette wheel selection the probability of being selected is proportional to an individual's fitness value. Therefore, highly fit individuals have a higher probability of being selected and hence of being represented in the next generation. In tournament selection a fraction of the individuals in the population are randomly selected into a subpopulation and competition is carried out to select the fittest individuals in each subpopulation. Table 2.1 and Figure 2.1 show a comparison between roulette wheel and tournament selection for a population of 10 individuals with randomly initialised fitness values between 0 and 100. The population is sampled 1000 times in each generation and the number of wins per individual is averaged over 10 generations and tabulated. For tournament selection, the subpopulation size is set to half of the population size.

Table 2.1 Comparison between roulette wheel and tournament selection mechanisms

	Fitness	Roulette wheel	Tournament
	9	15	0
	32	83	0
	40	70	0
	41	63	0
	48	69	2
	61	119	18
	62	96	56
	74	147	133
	75	139	304
	93	199	487
Total	535	1000	1000

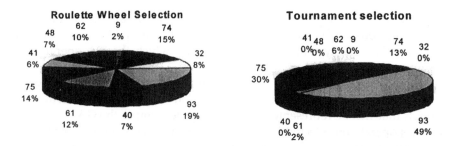

Figure 2.1 Pie charts comparing the different selection mechanisms

The results show that roulette wheel selection gives even the least fit members of the population a chance of being represented in the next generation. Tournament selection, on the other hand, is strongly biased in favour of the fittest individuals with a subset of the least fittest individuals guaranteed to disappear from the population in each generation.

Combination rules are used to introduce new individuals into the current population or to create a new population based on the current population. The genetic algorithm uses certain operators (genetic/search operators) [1,3] in the combination process. The most commonly used genetic operators are **crossover** and **mutation**. Other less commonly used ones include inversion and deletion. The combination rules act on individuals that have been previously selected by the selection mechanism. A reproduction process takes place between the selected individuals in the current generation to produce offspring that become individuals in the next generation (Figure 2.2).

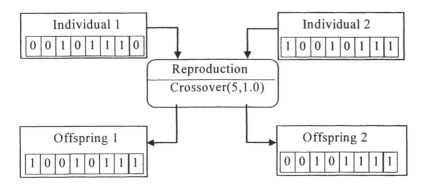

Figure 2.2 Schematic diagram depicting the reproduction process

Crossover is a combination rule that produces offspring in the Hamming interpolation of the parents. The crossover operator produces a non-uniform sampling probability distribution over the subspace. When standard or one-point crossover is applied on a pair of chromosomes, a real value generated at random is compared with the crossover threshold. If the value is less than the threshold, then a random site is picked for the crossover of genetic material between the two chromosomes. A number of alternative crossover strategies have also been suggested [4]. These include two-point crossover, multi-point crossover and uniform crossover. In two-point crossover, two sites are selected at random and the participating chromosomes swap genetic material between the sites. Multi-point crossover is similar to two-point crossover. Here, the number of crossover points is selected at random and alternative sections of genetic material are swapped between participating chromosomes. Finally, with uniform crossover, a crossover mask is generated initially for all chromosomes. For each mask position, a real value is generated at random and compared with the crossover threshold. If the value is less than the threshold, then the mask at the position is set, otherwise the mask is cleared. For a pair of participating chromosomes, exchange of genetic material takes place only at positions where the mask is set. Figures 2.3, 2.4 and 2.5 show how the different crossover strategies are applied to a pair of participating chromosomes and the resulting offspring.

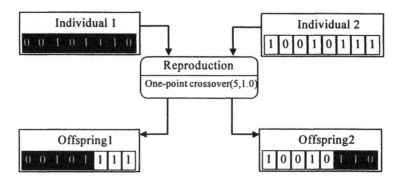

Figure 2.3 One-point crossover

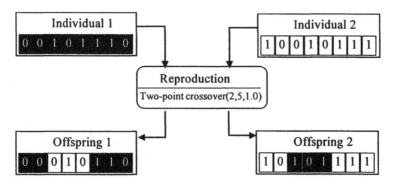

Figure 2.4 Two-point crossover

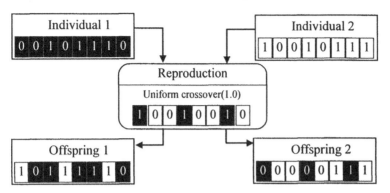

Figure 2.5   Uniform crossover

Mutation is the random occasional alteration of the information contained in the chromosome. The mutation probability determines how often mutation occurs. By itself, mutation is a random walk through the Hamming space, but when mutation is combined with crossover, it acts to improve the performance of the GA by preventing premature loss of genetic information from the GA population.

2.3   OBJECT-ORIENTED ANALYSIS OF GAS

An object-oriented analysis of GAs is performed to:

- understand the problem.
- identify objects that will remain important in the life of the application, and
- identify the relationships between the different objects and the ways in which the objects interact.

We can then design, and subsequently construct, a correctly working, robust and easy to maintain GA system.

2.3.1   Identifying Objects in the GA Domain

The process of analysis begins by identifying the objects found in the GA problem domain. As mentioned before, the objects in the domain can be obtained by a careful analysis of the problem description combined with expert knowledge of the particular problem domain. In the absence of the latter, domain analysis can be carried out to identify the objects that experts in the domain perceive to be important to applications in the domain. This is done by studying previous applications, by talking to the experts and by reading the literature. In the GA domain, five main objects have been identified: genes, chromosomes, individuals, the population and the genetic algorithm (GA). A dictionary has been created for the objects so that their use in the description of the GAs is unambiguous.

2.3.2 The Data Dictionary

- **Genetic algorithm:** Robust search mechanism based on population genetics and natural selection. A GA object is an algorithmic object that evolves highly fit individuals that represent a parameter encoding of a problem to be solved.
- **Population:** The collection of individuals on which the GA acts.
- **Individual:** A structure representing a parameter coding of the problem to be solved.
- **Chromosome:** In biology, a chromosome is a locus of alleles that determines the phenotype of an individual organism. In a GA, a chromosome is a collection of genes that holds structural information about individuals found in the GA population.
- **Gene:** A gene or allele is the information contained in a single position or locus of a chromosome. It represents the lowest level coding of the parameters of the problem to be solved by the GA.

It should be stressed that analysis is done in the problem domain, and analysis domain objects are problem domain objects. The aim is to uncover important objects and relationships so that a correct GA system can be designed. A domain object model is used to describe the static structure of the system and the relationships between the domain objects. A class diagram representing the domain object model is shown in Figure 2.6.

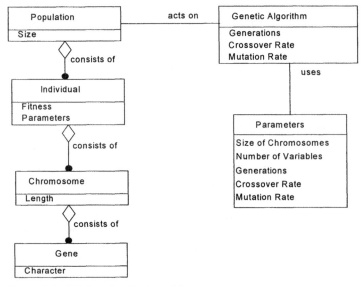

Figure 2.6 GA domain object model

The domain object model shows that a GA system makes use of a population of individuals. Each individual is composed of chromosomes and has a fitness value. Chromosomes in turn are a certain length and are made up of genes. Each gene has a character that represents the information that the gene encodes.

2.3.3   Discovering Object Operations

A dynamic model of the GA can be created to show the different states that a GA can be in and the events that it responds to. The behaviour of the GA is dependent on the operations that can be performed on it and the events that it can respond to. The dynamic model enumerates the important states that need to be captured in any GA implementation and hence the operations that are required in the GA interface. Dynamic models are expressed in terms of state-transition diagrams. Figure 2.7 shows a state-transition diagram for the GA.

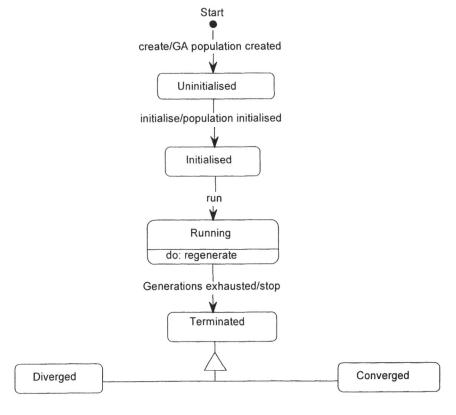

Figure 2.7   GA state-transition diagram

The different events, *create, initialise* and *run*, will cause a state transition in the GA when they are received. The action parts of the event (where present) will be carried out as a result of the state transition. For example, when a *create* event is received, the GA population is created and the state of the GA changes to *un-initialised*. On receipt of an *initialise* event, the population is initialised and causes a state transition to *initialise*. The *running* state is entered when the *run* event is received. While in the *running* state, the *regenerate* activity takes place to continually regenerate the GA population. When the activity finishes, the state automatically changes to *terminated*. The transition to state *terminated* occurs when the event *generations exhausted* is received. The states *converged* and *diverged* are substates or specialisations of *terminated*.

2.4   OBJECT-ORIENTED GA DESIGN

The models developed in the analysis phase are used as inputs to the design phase. The aim of the design phase is to create a computer implementable blueprint for the GA system. During object-oriented design, the attributes of the domain objects are determined and the objects themselves are reorganised into class and inheritance hierarchies for easy and efficient implementation. New objects are also added to the analysis domain object model which allow the GA to be constructed efficiently and robustly. The new objects, sometimes described as system objects, are not found in the vocabulary of the problem but enable a modular and reusable system to be built. The design process will also seek to determine efficient data structures and the correct level of granularity for realising the domain objects and the relationships between them. Finally, the system is partitioned into subsystems which can be constructed and tested separately so that the overall complexity of the system is reduced further. The static structure of the system is given by the system object model shown in Figure 2.8. The design diagram shows a more detailed object model with the associated relationships. In the design, an alphabet object is used to represent genes in the classifier system domain. Alphabets can be a '1', '0' or '#', where '#' represents a 'don't care'. An array data structure is used to efficiently implement collections. Chromosomes and vectors are special collections of alphabets and numbers, respectively, and so inherit their behaviour from Arrays. Each Individual is made up of a Chromosome and a fitness value. The GA has a *uses* relationship with both the population and the parameter objects used for collecting GA and problem parameters such as population size, chromosome string size, number of generations, crossover and mutation probabilities, etc. from the users.

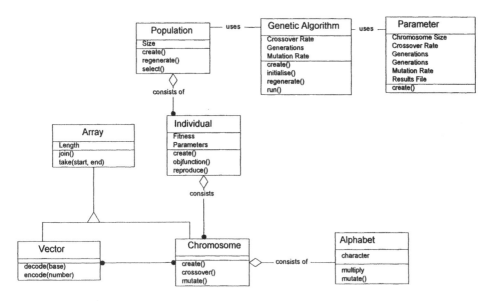

Figure 2.8 System object model of GA

2.4.1 Object Design

For the GA, the system design phase that will require the partitioning of the GA into subsystems is not necessary. It is not easy to visualise a partitioning of the GA into meaningful subsystems that simplify the implementation and can stand on their own. The input subsystem consists of a single parameter object that requests the GA and problem parameters from the user. The remaining portion constitutes what is traditionally regarded as a GA and so further partitioning is not necessary. Object design is carried out to fully specify the classes and relationships that have been identified so that a solution can be implemented. The objects can be fully specified by creating actual scenarios in which the objects will be used so as to determine the nature of the messages received by the different objects. Two scenarios, create, when the GA is created, and run, when the GA is running, are presented to show how the different objects interact. The scenarios are expressed in terms of object interaction diagrams. In the create scenario shown in Figure 2.9, a parameter is initialised with a request for the user to enter the system and problem parameters. A create() call to the GA object results in a create() call to the population object. The population in turn issues create() calls for each individual in the population. The creation of an individual object leads to a further create() call to the chromosome object. Finally, the chromosome calls create() for array and alphabet objects to complete the sequence of create() calls. This sequence results in the creation of the GA object at the highest level. In creating the interaction diagrams for a particular scenario, only the very important messages have been enumerated. Furthermore, the messages are at a high-level of abstraction so that the design diagrams can be kept simple and focused. Each high-level message subsequently can be refined into a number of lower level messages which can be depicted as separate scenarios or use cases. At the lowest level, the messages are simply member functions for the public interface of objects.

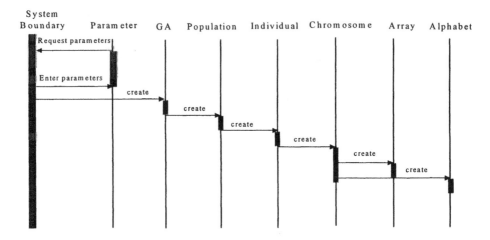

Figure 2.9 Object interaction diagram for the create scenario

The running of the GA can be described by run scenario expressed in Figure 2.10.

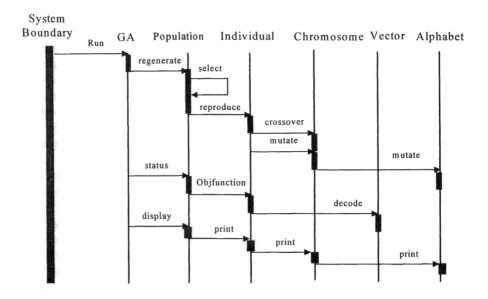

Figure 2.10 Object interaction diagram for the run scenario

When the GA is first run, a regenerate() message is sent to the population object to create a new population from the current population. Population issues select() messages to itself to obtain individuals for reproduction and sends a reproduce() message to individual. Individual in turn sends crossover() and mutate() messages to chromosome. To honour any mutation requests, Chromosome has to send a mutate() request to Alphabet, which causes a random change in the Alphabet. When the population has been regenerated, a status() message is sent from GA to Population to perform the housekeeping. Population then sends an Objfunction() message to Individual to update the fitness values of the new individuals in the population. Individual issues a decode() message to Vector so that the values of their Chromosomes can be decoded as binary vectors. Finally, the results of each generation are displayed from a sequence of display() messages sent between the objects. The interaction diagrams show how the Genetic Algorithm objects co-operate at a lower level to realise the required high level behaviour. The diagrams also show the important messages that each object has to respond to and hence the nature of the member functions required in the object's public interface. Other member functions are also added to the object's interface to support construction, destruction and general housekeeping activities like copying, storage and retrieval for each object. Fully specified structures for Alphabet, Individual and GA objects are shown in Figure 2.11.

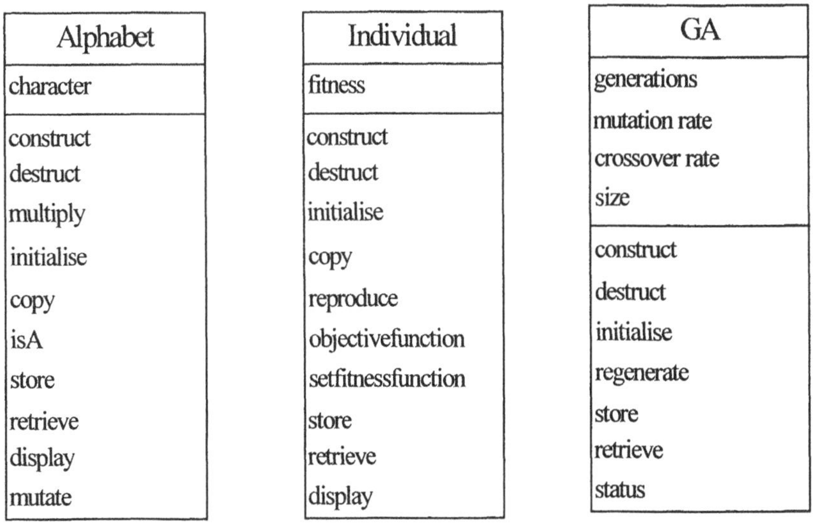

Alphabet	Individual	GA
character	fitness	generations
		mutation rate
construct	construct	crossover rate
destruct	destruct	size
multiply	initialise	
initialise	copy	construct
copy	reproduce	destruct
isA	objectivefunction	initialise
store	setfitnessfunction	regenerate
retrieve	store	store
display	retrieve	retrieve
mutate	display	status

Figure 2.11 Fully specified object structures for some GA objects

2.4.2 Implementation

Other GA objects have been similarly realised. The C++ language also provides direct support and language constructs for implementing the relationships between the objects. Composition/Aggregation relationships such as that between Population and Individual are realised using pointers and references while inheritance hierarchies are used to implement Generalisation/Specialisation relationships. For example, the object diagram showed that an Array is a general data structure used to implement collections, while Chromosomes and Vectors are special cases of Arrays. The relationship between Arrays on the one hand and Vectors and Chromosomes on the other is said to be a Generalisation/Specialisation relationship. As a result of the inheritance relationship, data and functions declared in Array are reused in Vectors and Chromosomes without the need for redefinition. Functionality such as crossover(), which is specific to Chromosomes, and decode(), which is specific to Vectors, are then added to Chromosomes and Vectors, respectively. Functions that have been defined in the Array class can be refined in the Vector and Chromosome classes to be more specific or more efficient, or both. The Array class is called the Base or superclass while the Chromosome and Vector classes are known as Derived or subclasses.

2.4.3 Object-Oriented Testing

Unit testing is carried out on each GA object to ensure satisfactory behaviour. The set of messages that the object needs to respond to, can be simulated and sent to an object independent of the other objects in the GA. Closely coupled objects and objects in inheritance relationships such as Arrays, Chromosomes and Vectors in the GA are tested together. The test process is thus simplified and predictable. For example, the test result for Alphabet is shown in Figure 2.12.

Figure 2.12 shows the behaviour of the Alphabet object when the different messages in its interface are issued. Two strings of alphabet are created initially at random. Then the strings are initialised with a user-supplied string of characters stored in the test file. Mutation, Multiplication, Storage and Retrieval tests are then carried out. The results confirm that the Alphabet object behaves according to the required specification. As can also be seen, low rates of mutation cause little or no change in the original strings. As the rate of mutation increases, the discrepancies between the original strings and mutated strings also increase confirming that the mutation operation performs satisfactorily.

After unit testing, integration testing is carried out. All the finished classes are put together as an application to check its operation on real problems and to remove subtle mistakes. Because of the flexible architecture, the GA implementation can be configured to solve a wide variety of searching problems. Options have been included to set the number of objective function parameters, the crossover strategy, the population replacement strategy and even the nature of the selection function for selecting fit individuals from the GA population. Furthermore, the size of the encoding for the parameters in an optimisation problem can be varied to improve the resolution of the search.

The following example shows how the GA is used to search for an optimal value in a space where the Gaussian mixture objective function consists of two Gaussian functions. They have similar means but one has one-tenth the amplitude and five times the standard deviation of the other one. The objective function is expressed as shown in Equation (1) below:

$$f(x,y) = 0.9 \times \frac{1}{\sqrt{2\pi\sigma_1^{\,2}}} \exp\!\left(-\frac{(x-\mu_1)^2}{\sigma_1^{\,2}}\right) \times \frac{1}{\sqrt{2\pi\sigma_2^{\,2}}} \exp\!\left(-\frac{(y-\mu_2)^2}{\sigma_2^{\,2}}\right)$$
$$+ 0.1 \times \frac{1}{\sqrt{2\pi(5\sigma_1)^2}} \exp\!\left(-\frac{(x-\mu_1)^2}{(5\sigma_1)^2}\right) \times \frac{1}{\sqrt{2\pi(5\sigma_2)^2}} \exp\!\left(-\frac{(y-\mu_2)^2}{(5\sigma_2)^2}\right) \tag{1}$$

where μ_1 and σ_1 are the mean and variance, of the first function, μ_2 and σ_2 are the mean and variance of the second one.

```
Initial display of both Alphabets
0 0 0 # 1 # # 1 1 1
1 0 # 1 0 1 # # # #
Alphabets after initialisation
Original I : 1 1 0 0 0 0 1 0 1 0
Original II: 1 # 1 0 0 1 # # # 0
Mutation test
Mutation Prob    Original I              Original II
                1 1 0 0 0 0 1 0 1 0      1 # 1 0 0 1 # # # 0
   0.001        1 1 0 0 0 0 1 0 1 0      1 # 1 0 0 1 # # # 0
   0.005        1 1 0 0 0 0 1 0 1 0      1 # 1 0 0 1 # # # 0
   0.009        1 1 0 0 0 0 1 0 1 0      1 # 1 0 0 1 # # # 0
   0.01         1 1 0 0 0 0 1 0 1 0      1 # 1 0 0 1 # # # 0
   0.05         1 1 0 0 0 1 1 0 1 0      1 # 1 0 0 1 # # # 0
   0.09         1 1 0 0 0 1 1 0 1 0      1 # # 0 0 1 # # # 0
   0.1          1 1 0 0 0 1 1 0 1 0      1 # # 0 0 1 # # # 0
   0.5          1 # # # 0 1 1 0 1 #      1 0 # 0 1 1 # # # 0
   0.9          0 # 1 # # # 0 # # #      0 1 1 # 1 1 1 # 0 #
   1            1 0 0 # 0 1 0 # 1 0      # # 1 1 1 # 0 0 0 1
Multiplication test!
1 0 0 # 0 1 0 # 1 0
# # 1 1 1 # 0 0 0 1
 Result :       1 1 0 1 0 1 1 1 0 0
testing file input /output
aArray, bArray, cArray have been written to file alfa.dat
Reading test from file
The alphabets read are
1 0 0 # 0 1 0 # 1 0
# # 1 1 1 # 0 0 0 1
1 1 0 1 0 1 1 1 0 0
```

Figure 2.12 Results of testing the GA Alphabet class

A plot of the objective function is shown in Figure 2.13. The GA is configured to search the two-dimensional space represented by the objective function for the maximum value. The aim of the search is to find two real values x and y in the range 0 to 5 representing the co-ordinates of the maximum point of the objective function. Figure 2.14 shows how the objective function behaves in the vicinity of the maximum. A numerical solution using Mathematica shows that the function has a single maximum at 6.39449 when $x = 2$ and $y = 3$. The objective function then falls off very rapidly to zero all around the maximum.

In the GA, both x and y are encoded as 20-bit strings which are then concatenated to form a 40-bit chromosome for each individual. To calculate the individual's fitness value, each chromosome is split into its two constituent parts which are then decoded and scaled according to Equation (2) into the x and y values. The values obtained are evaluated by the Gaussian mixture function to obtain the fitness of the individual:

$$x = y = \left(\frac{decoded\_value}{2^{20} - 1} \right) \times 5 \tag{2}$$

Gaussian Mixture Objective function

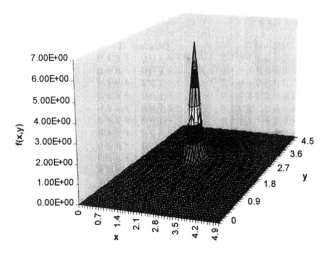

Figure 2.13 Objective function plot for $\mu_1 = 3$, $\mu_2 = 2$, and $\sigma_1 = \sigma_2 = 0.15$

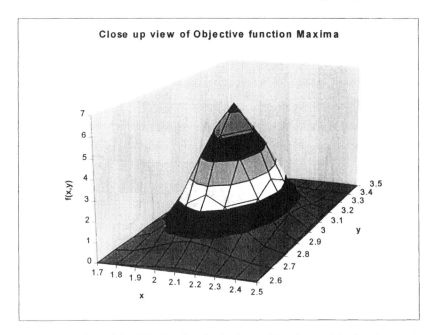

Figure 2.14 Behaviour of the objective function in the neighbourhood of the function maximum.

The GA is initialised at random with a population of 30 individuals. The number of generations is set to 30 and the GA is then run a number of times with different values for the GA parameter, i.e. mutation rate, crossover rate and population replacement strategy. As with the previous objective function, the results are compared with a random population search of the function space where the crossover rate is set to 0, the mutation is set to 1 and a random selection strategy is used. The population replacement strategy can be either Generational or Elitist. Figure 2.15 shows how the maximum and average fitnesses vary when a random population search is used in conjunction with Generational replacement. Since Generational replacement completely replaces the GA population in each generation and because no information is passed from one generation to another, the search will never converge. In Figure 2.16 a random population search is carried out in conjunction with Elitist population replacement, i.e. the best individual in each generation is retained in the next generation. The figure shows that the random population search can still arrive at an optimal or near optimal solution if the best individual in each generation is retained, but the variation of total and average fitnesses on a per generation basis is still very erratic.

Fitness Vs Generation

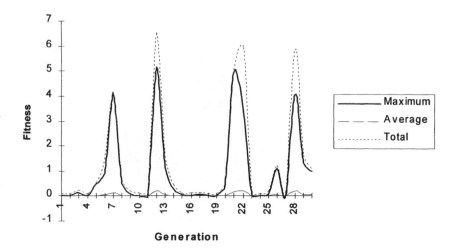

Figure 2.15   Random search with Generational replacement

Fitness vs Generation

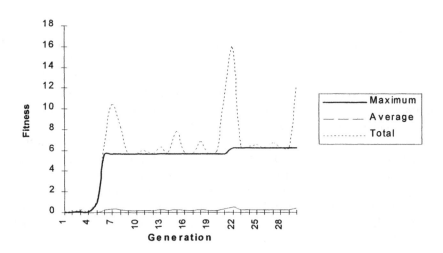

Figure 2.16   Random search with Elitism

Figures 2.17 to 2.19 show the equivalent graphs for GA search. The convergence of the fitness and x and y values with generation are also tabulated in Table

2.2. In the graphs, the total fitness has been omitted so that the maximum and average fitnesses are visible.

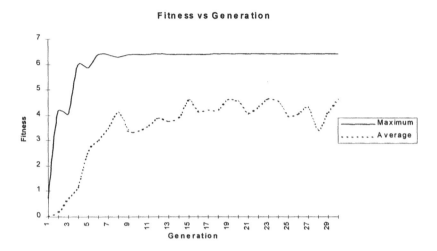

Figure 2.17 Variation of maximum and average fitness values with generation (crossover = 0.8, mutation = 0.1, Roulette selection, two-point crossover and Generational replacement)

Table 2.2 Best of generation fitness and (x, y) values vs. generation number for GA search

Gen	Figure 2.17	Figure 2.18	Figure 2.19	Figure 2.19 (x)	Figure 2.19 (y)
1	0.728291	0.025201	0.506987	2.4353	1.89206
2	4.17134	0.027658	1.08185	3.3728	1.89084
3	4.04754	2.20231	1.68198	2.79052	1.89077
4	5.97665	2.54161	3.18862	3.13842	1.89046
5	5.87153	4.6623	4.83984	3.13842	1.89298
6	6.36692	5.51842	5.0433	3.1042	2.0482
7	6.37606	5.66355	5.56414	3.09444	2.04091
8	6.26938	6.19981	5.78558	3.09444	2.0471
9	6.3662	6.34124	5.83347	3.02551	1.9711
10	6.36927	6.34922	6.00828	3.02551	1.96775
11	6.37617	6.35054	6.33573	3.02427	1.96716
12	6.39095	6.38343	6.28355	3.02512	1.96653
13	6.38138	6.38343	6.29152	3.02591	1.96896

Table 2.2 continued

14	6.38314	6.39276	5.89751	3.02608	1.96896
15	6.38754	6.39407	6.04052	2.98702	1.96896
16	6.38754	6.39407	6.04052	3.0228	1.98669
17	6.38478	6.39445	5.8	2.98642	1.98669
18	6.39216	6.39419	5.8	2.98642	1.98648
19	6.39137	6.39436	6.05812	2.98642	1.98648
20	6.39371	6.39438	5.98166	2.98642	1.98667
21	6.39244	6.39438	6.00832	2.98642	1.98632
22	6.3937	6.39438	5.88713	3.01687	1.98667
23	6.39369	6.39447	6.09555	3.0261	1.98423
24	6.39373	6.39447	5.91747	3.01503	2.00948
25	6.39253	6.39448	6.15635	3.01621	2.01714
26	6.39337	6.39447	5.98371	3.01621	2.01714
27	6.39323	6.39446	6.12896	3.01507	2.00204
28	6.3938	6.39386	5.99622	3.01684	1.99147
29	6.39382	6.39447	6.17239	3.02549	2.00296
30	6.39386	6.39447	6.20304	2.98658	2.00376

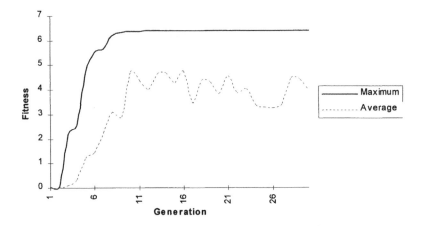

Figure 2.18 Variation of maximum and average fitness values with generation (crossover = 0.8, mutation = 0.1, Tournament selection, two-point crossover and Generational replacement)

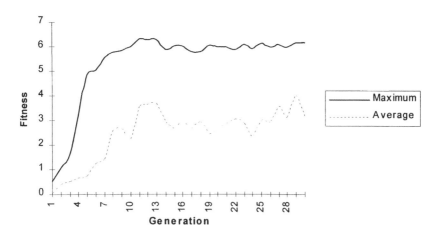

Figure 2.19 Variation of maximum and average fitness values with generation (crossover = 0.8, mutation = 0.1, Tournament selection, one-point crossover and Generational replacement)

The results show that a random search with Generational replacement has a very small probability of arriving at an optimal solution at the end of 30 generations. This is because no information is stored between generations and the search results in the thirtieth generation are just as likely as those in the first generation. On the other hand, when Elitism is added to the random search, the best individual in each generation is retained and there is thus a significant performance improvement in the search process. However, the chances of arriving at an optimal solution within 30 generations is still very small. For a search resolution accurate to two decimal places, the search space contains 250000 points. If the resolution is increased to three decimal places, then the cardinality of the search space increases to 25 million points. Also, if either the search range or the search resolution is increased, then the number of points rapidly becomes intractable. It becomes almost impossible for a brute force search method such as random search to locate a single point in such large search spaces. A GA search, on the other hand, converges to an optimal or near optimal solution in a very short time.

The results show how the genetic search converges to an optimal solution in 30 generations for three different runs of the GA. The table also shows how the decoded values for the x and y co-ordinates of the maximum vary with generation for the third run of the GA. In Figure 2.17 the GA converges to a final value of 6.39386. This value is identical, to three decimal places, to the analytically derived optimum value and is just 0.00063 or 0.000985% different from the maximum. In the second run, shown in Figure 2.18, the search results are even more impressive. The combination of Tournament selection, two-point crossover and Generational replacement causes the GA to converge to within 0.00002 of the required optimum value. In the third run, shown in Figure 2.19, Tournament selection combined with one-point crossover and Generational replacement causes a convergence to a suboptimal value. Even here, the converged GA is within 2.1% of the required value and can be considered as near optimal for most purposes. The optimality of the solution can be appreciated when the variation in the best of the Generation parameters is examined. The optimal values for x and y are 2 and 3, respectively. As shown in Figure 2.20, the error in the decoded values for x and y are relatively large at the start of the GA run. As the GA converges, the values tend to stabilise with very little fluctuation around the optimal values.

The above examples demonstrate that the GA implementation is capable of successfully searching higher dimensional spaces in the solution of complex optimisation problems. In subsequent chapters of this book where GA is used to search for a solution, the uniform crossover is adopted.

Best of Generation paramerters vs Generation

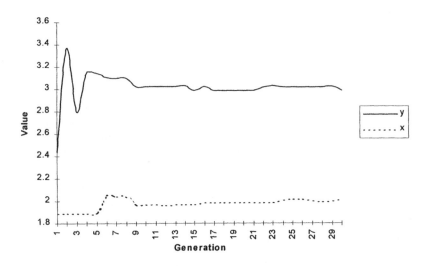

Figure 2.20   Variation of decoded parameters with Generation

2.5   EVOLUTIONARY PROGRAMMING (EP)

EP is a computational intelligence method in which an optimisation algorithm is the main engine for the process of three steps; namely, natural selection, mutation and competition. According to the problem, each step could be modified and configured in order to achieve the optimum result. It is a stochastic optimisation strategy which places emphasis on the behavioural linkage between parents and their offspring, rather than seeking to emulate specific genetic operators as in GAs [5,6]. Crossover in genetic algorithms often destroys the essential behavioural link between each parent and its offspring. However, mutation ensures the functionality of the next generation. Therefore EP tends to generate more effective and efficient searches. Such arguments have also been made in [7] and [8]. As such there is considerable evidence that EP provides a more robust method for solving constrained optimisation problems than do GAs [5]. Combinatorial and real-valued function optimisation in which the optimisation surface possesses many locally optimal solutions are well suited for EP.

2.5.1 Features of EP

EP belongs to the class of population-based search strategies. They operate on populations of real values (floating points) that represent the parameter set of the problem to be solved over some finite ranges. Each representation is an individual in the EP population. The population is initialised with random individuals at the start of the EP run. The EP searches the space of possible real values for better individuals. The search is guided by fitness values returned by the environment. This gives a measure of how well adapted each individual is in terms of solving the problem and hence determines its probability of appearing in future generations. Two types of rules are used by EP in its search for highly fit individuals, namely the selection rule and combination rule. The selection rule is used to determine the individuals that will be represented in the next generation. It includes Competition in which each individual in the combined population has to compete with some other individuals to get chance to be transcribed to the next generation. The combination rule operates on selected individuals to produce new individuals that appear in the next generation. The selection mechanism is based on a fitness measure or objective function values, defined on each individual in the population. The combination rule is used to introduce new individuals into the current population or to create a new population based on the current population. EP uses only one operator in the combination process. The most commonly used evolutionary operator is mutation. Mutation is the random occasional alteration of the information contained in the individual. The combination rule acts on individuals that have been previously selected by the selection mechanism.

A number of selection and combination methods have been reported in different research papers but no standard has been set up so far. Therefore, two separate methods in the form of two algorithms are presented here to clarify the issue. Both methods have been tested and used and are presented in the following sections.

2.5.2 Two Simple Algorithms for EP

EP considers the problem of minimising a function $f(\alpha)$, where α is a vector of real values. Two implementation methods are presented in the following sections. The first is classical EP and the second is adaptive EP, which is described with a more detailed presentation.

2.5.2.1 Classical EP

Classical EP can be implemented as follows:

1. Generate n random vectors $\alpha_1, \alpha_2, ..., \alpha_n$ uniformly distributed on [R1,R2]
2. Loop until finished
    2.1 Sort $\alpha_1, \alpha_2, ..., \alpha_n$ by $f(\alpha_i)$ in descending order
    2.2 Delete the bottom half of α_i.
    2.3 Replace bottom half by $\alpha_i + \eta_i, i = 1, ..., n/2$, $\eta_i \approx N(0, \sigma)$

where $N(0, \sigma)$ is uniformly distributed with mean 0 and variance σ. Typically, the random mutation η_i consists of samples from a multidimensional normal distribution with small variance. There are many variations of this classical EP. For example, the σ, which is the mutation parameter, can be either constant or adaptive. This will be discussed when EP and ANN are combined in the next chapter.

2.5.2.2 Adaptive EP with β

Another version of EP implementation is introduced as follows:

- **Initialisation:** The initial population consists of individuals and is created randomly. The fitness score f_i of each individual p_i is obtained by a fitness function.
- **Statistics:** The maximum fitness, minimum fitness, sum of fitness and average fitness of this generation are calculated.
- **Mutation:** Each p_i is mutated in order to generate a new population. The new individuals, p_{i+m}, are calculated in accordance with the following equation:

$$p_{i+m,j} = p_{i,j} + N\left(0, \beta(x_{j\max} - x_{j\min}) \frac{f_i}{f_{\max}} \right), \quad j = 1, 2, ..., n \qquad (3)$$

where $p_{i,j}$ denotes the jth element of the ith individual; $N(\mu, \sigma^2)$ again represents a Gaussian random variable with mean μ and variance σ^2; $f_{\max}$ is the maximum fitness of the old generation, which is obtained in Statistics; x_{jmax} and x_{jmin} are the maximum and minimum limits of the jth element; β is the mutation scale that is given as $0 < \beta \leq 1$. If any $p_{i+m,j}$, $j=1, 2, ..., n$, where n is the number of control variables, exceeds its limit, $p_{i+m,j}$ will be given with the limit value. The mutation scale should be changed to prevent a search being trapped in a local minimum. An adaptive mutation scale is given by changing β after each mutation. The mutation scale of each mutation is changed according to the following:

$$\beta(k+1) = \begin{cases} \beta(k) - \beta_{step} & f_{min}(k) \text{ unchange} \\ \beta(k) & f_{min}(k) \text{ decrease} \\ \beta_{final} & \beta(k) - \beta_{step} < \beta_{final} \end{cases} \quad (4)$$

The corresponding fitness f_{i+m} is obtained by using the fitness function, as presented above, for p_{i+m}. A combined population is formed with the old generation and the mutated old generation with $2m$ individuals. The initial β is 1 then it decreases by β_{step} which is set from 0.001 to 0.01. β_{final} is set to 0.005. β values depend on the number of generations and the complexity of the system.

- **Competition:** Each individual p_i in the combined population has to compete with some other individuals to get chance to be transcribed to the next generation. A weight value w_i is assigned to the individual according to the competition as follows:

$$w_i = \sum_{t=1}^{q} w_t \quad (5)$$

where q is the competition number, which is largely dependent on the parameters of the system such as population size. w_t is a number of $\{0, 1\}$, which represents win 1, or loss 0, as p_i competes with a randomly selected individual p_r in the combined population. w_t is given by the following equation:

$$w_t = \begin{cases} 1 & f_i > f_r \\ 0 & \text{otherwise} \end{cases} \quad (6)$$

where f_r is the fitness of a randomly selected individual p_r and f_i is the fitness of p_i. When all individuals p_i, $i=1, 2, ..., 2m$, get their competition weights, they will be ranked in descending order of their corresponding value w_i by a sorting algorithm. The first m individuals are transcribed along with their corresponding fitness f_i to be the basis for the next generation.

2.6 OBJECT-ORIENTED ANALYSIS, DESIGN AND IMPLEMENTATION OF EP

2.6.1 Object-Oriented Analysis of EP

An object-oriented analysis of EP is performed so that a correctly working, robust and easy to maintain EP system can be designed and subsequently constructed.

In the EP domain, four main objects have been identified: Vectors, Individuals, Population and EP. A dictionary has been created for the objects so that their use in the description of the EP is unambiguous.

2.6.2 The Data Dictionary

- **EP:** A robust search mechanism that is based on population evolution and natural selection. An EP object is an algorithmic object that evolves highly fit individuals that represent parameters to solve a problem.
- **Population:** The collection of individuals on which the EP acts.
- **Individual:** A structure represents a parameter coding of the problem to be solved.
- **Vector:** A collection of real type values that holds structural information about individuals found in the EP Population.

A class diagram representing the domain object model is shown in Figure 2.21.

The domain object model shows that an EP system makes use of a Population of Individuals. Each Individual is composed of a Vector and has a fitness value. Vectors in turn are of a certain Length and are made up of certain precise real numbers.

2.6.3 Developing Object Operations

A dynamic model for EP can be created to show the different states that the EP can be in and the events to which it responds. The behaviour of the EP is dependent upon the operations that can be performed on it and the events to which it can respond. The dynamic model, enumerates the important states that need to be captured in any EP implementation and hence the operations required in the EP interface. Dynamic models are expressed in terms of state-transition diagrams. Figure 2.22 shows a state-transition diagram for the EP.

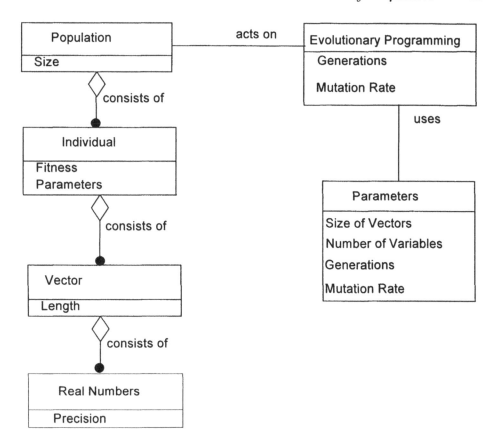

Figure 2.21 EP domain object model

2.6.4 Object-Oriented EP Design

The static structure of the system is given by the system object model shown in Figure 2.23. The design diagram shows a more detailed object model with the associated relationships. In the design, floating point or double precision numbers are used to represent real number values in the system domain. A Vector data structure is used to implement collections efficiently. Individuals are special collections of Vectors so that they inherit their behaviour. Each Individual is made up of a Vector and a fitness value. The EP has a *uses* relationship with both the Population and the Parameter object used for collecting EP and problem parameters such as Population size, Vector size, number of generations, mutation parameters, etc. from the users.

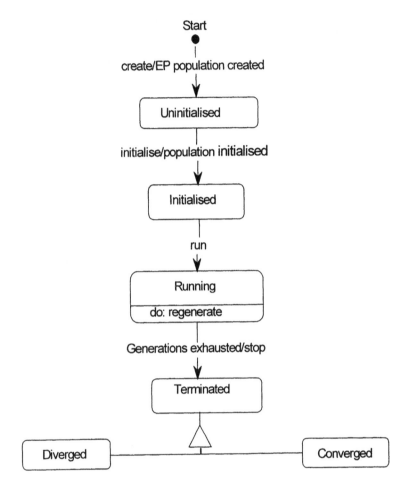

Figure 2.22 EP state-transition diagram

The explanation for Figure 2.22 is similar to that given for Figure 2.7.

2.6.5 Object Design

For EP, object interaction diagrams for the create and run scenarios are shown in Figures 2.24 and 2.25. These could be explained in a similar way as for GA.

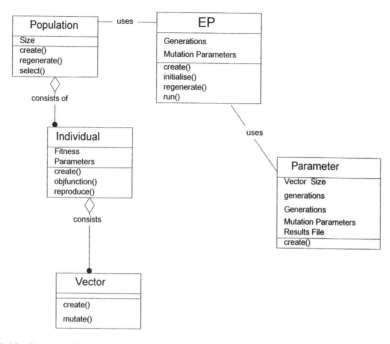

Figure 2.23 System object model of EP

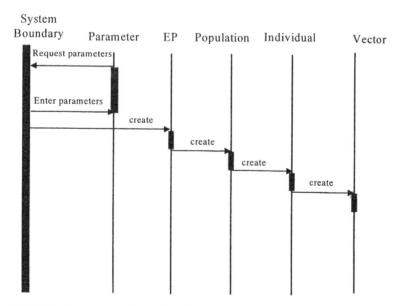

Figure 2.24 Object interaction diagram for the create scenario

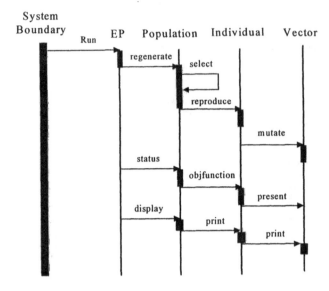

Figure 2.25 Object interaction diagram for the run scenario

Fully specified structures for Vector, Individual and Evolutionary Programming objects are shown in Figure 2.26.

Vector	Individual	EP
real number	fitness	generations
		mutation parameters
construct	construct	
destruct	destruct	size
multiply	initialise	
initialise	copy	construct
copy	reproduce	destruct
isA	objectivefunction	initialise
store	setfitnessfunction	regenerate
retrieve	store	store
display	retrieve	retrieve
mutate	display	status

Figure 2.26 Fully specified object structures for some EP objects

2.6.6 Implementation

The finished design could be implemented with the C++ programming language. There is a one-to-one translation between the objects in the design object model and classes.

2.6.7 Object-Oriented Testing

As in the the GA case, EP object testing is carried out to ensure satisfactory behaviour. Comparisons between EP and GA for the same problems have been reported in [5] and also in some subsequent chapters, comparisons between these two approaches will be given for example, in Chapter 4. Chapters 4 to 11 show EP applications in power engineering.

2.7 REFERENCES

[1] D Goldberg, *Genetic Algorithms in Search, Optimisation and Machine Learning*, Addison-Wesley Publishing Company, 1989.
[2] L Davis (Editor), *Handbook of Genetic Algorithms*, Van Nostrand Reinhold, 1991.
[3] A Ackley, *A Connectionist Machine for Genetic Hill Climbing*, Kluwer Academic Publishers, 1987.
[4] H Adeli and S Hung, *Machine Learning: Neural Networks, Genetic Algorithms and Fuzzy Systems*, John Wiley & Sons, 1995.
[5] D B Fogel, 'A comparison of evolutionary programming and genetic algorithms on selected constrained optimization problems', *Simulation*, **64**, June 1995, 397-404.
[6] T Bäck and H P Schwefel, 'An overview of evolutionary algorithms for parameter optimization', *Evolutionary Computation*, **1**, 1993, 1-24.
[7] P J Angeline, G M Saunders and J B Pollack, 'An evolutionary algorithm that constructs recurrent neural networks', *IEEE Transactions on Neural Networks*, **5**, 1994, 54-65.
[8] W Atmar, 'Notes on the simulation of evolution', *IEEE Transactions on Neural Networks*, **5**, 1994, 130-148.

3

Hybrid Evolutionary Algorithms and Artificial Neural Networks

In this chapter, different hybrid evolutionary algorithms and artificial neural networks are presented. The evolution of connection weights, architectures and training algorithms are introduced and each approach is described and analysed. Critical issues related to different object-oriented integration between developed ANN, EP and GA objects are discussed.

3.1 INTRODUCTION

Research on potential interactions between connectionist learning systems, such as artificial neural networks (ANNs), and evolutionary search procedures, such as evolutionary programming (EP) and genetic algorithms (GAs) has attracted a lot of attention recently. Evolutionary ANNs (EANNs) can be considered as the combination of ANNs and evolutionary search procedures. The interest in EANNs has been growing rapidly in recent years, as this research not only advances our understanding of adaptive processes in nature, but also helps computer scientists and engineers to develop more powerful intelligent systems. In this chapter, first, the principles and a number of different methods of evolution's in EANNs are presented and distinguished, for example the evolution of connection weights, architectures and training algorithms. Each approach is described and analysed. Critical issues related to different evaluations are also discussed. We place emphasis on the evolutionary programming and artificial neural networks EP-ANN hybrid system, since there has been very little research work carried out in this particular combination of EANNs [1]. The combination of GAs and ANNs, GA-ANN, is also presented, but with less emphasis than EP-ANN because there are a number of research reports available on GA-ANN. The interactions between developed ANN, EP and GA components are presented in order to explore possible benefits arising from these combinations. Instead of using them individually, the main emphasis is the object-oriented integration of the techniques and its aspects. The object-oriented technique gives us the ability to combine the existing developed objects and create new components. In order to perform this task, a thorough analysis of both objects should take place. This includes an understanding of the principles of the hybrid system, identifying the objects and the relationships between them, and the ways in which the objects interact.

3.2 WHY DEVELOP EANNS?

Artificial neural networks (ANNs) offer an attractive paradigm for the design and analysis of adaptive, intelligent systems for a broad range of applications in artificial intelligence and cognitive modelling for a number of reasons, including:

- resilience to the failure of components;
- robustness in the presence of noise;
- amenability to adaptation and learning (by the modification of computational structures employed);
- potential for massively parallel computation; and
- resemblance to biological neural networks (brains).

Despite much research activity, which has led to the discovery of several significant theoretical and empirical results and applications, the design of ANNs for specific applications to a large extent, relies mostly on past experience. Furthermore, the performance (and cost) of ANNs for particular problems is critically dependent on the choice of primitives (neurons or processing elements), network architecture and the training algorithm(s) used. For example, many of the popular training algorithms used in ANNs essentially search for a suitable setting of modifiable parameters, such as weights and biases. Clearly, in order for this approach to succeed, the desired setting of parameters must, in fact, exist within the space being searched and the search algorithm used must be able to find it. Even when a suitable setting of parameters can be found, the ability of the resulting network to generalise on unseen data or the cost of the hardware realisation or the cost of using it may be far from optimal. This is a very critical issue. These factors make the process of ANN design difficult. In addition, the lack of sound design principles constitutes a major hurdle in the development of large-scale ANNs for a wide variety of practical problems. Thus, techniques for automating the design of neural architectures for particular classes of problems under a wide variety of design and performance constraints are clearly of interest. Motivated by this, some researchers have recently begun to investigate constructive or generative ANN training algorithms that extend the search for the desired input/output mapping to the space of appropriately constrained network topologies by incrementally constructing the required network. Against this background, a logical next step is the exploration of more powerful techniques for efficiently searching the space of network architectures. Thus, the design of neural architectures presents us with a challenging multi-criterion optimisation problem. Evolutionary algorithms offer an attractive and relatively efficient, randomised opportunistic approach to search for global or near global optimal solutions for a variety of problem domains. The design of efficient neural architectures for specific classes of problems under given sets of design and performance constraints is therefore a natural candidate for the application of evolutionary algorithms. In addition to the development of new and powerful techniques for neural engineering, the exploration of evolutionary approaches to the design of neural architectures is likely to shed light on important open problems in artificial intelligence and cognitive modelling. What distinguishes evolutionary design of neural architectures from general methods for program synthesis is the commitment to a particular class of computing structures; namely, massively parallel, highly interconnected networks of relatively simple processing elements as opposed to, say, LISP programs. Given the equivalence of several computation models, including lambda calculus of LISP programs and Turing machines, the choice of any given model is mostly a matter of convenience. Some models are better suited to express certain classes of computations than others and may have some cost/performance advantages over others with particular implementation media. ANNs represent a particularly attractive computation model for a broad range of problems in artificial intelligence and cognitive modelling. Furthermore, the study of designs for neural architectures provides a useful research strategy for exploring answers to many fundamental questions of interest in neuroscience. For example: How does the brain get wired or programmed through training or evolution to obtain a solution?

3.3   EANN METHODS

Three general kinds of evolution can be performed on ANNs, including the evolution of :

- connection weights and biases;
- architectures; and
- training algorithms.

These are the levels at which evolutionary search procedures come into EANNs. The first two are emphasised here by two EANN systems; namely, EP-ANN and GA-ANN.

3.4   THE EVOLUTION OF EANN WEIGHTS

Supervised training has mostly been formulated as a weight training process. Effort is made to find an optimal (near optimal) set of connection weights for a network according to some criteria. The total mean square error (MSE) between the actual output and the target output is used to guide BP search in the weight space. There have been some successful applications of BP algorithms in various areas. However, drawbacks with the BP algorithm do exist owing to its gradient descent nature. It often gets trapped in a local minimum of the error function and is very inefficient in searching for the global minimum of a function which is vast, multimodal and non-differentiable. One way to overcome the shortcomings of BP as well as other gradient descent search-based training algorithms, is to consider the training process as the evolution of connection weights towards an optimal (near optimal) set defined by a fitness function. From such a point of view, global search procedures like EP or GAs can be used effectively to train an EANN. The fitness of an EANN can be defined by the aforementioned total MSE.

The evolutionary training approach is divided into two major steps. The first is to decide the representation scheme of connection weights, e.g. binary strings and real values. The second step is the evolution itself driven by the EA. Different representation schemes and EAs can lead to quite different training performance in terms of training time and accuracy. A typical evolution cycle of connection weights with EP is shown in Figure 3.1.

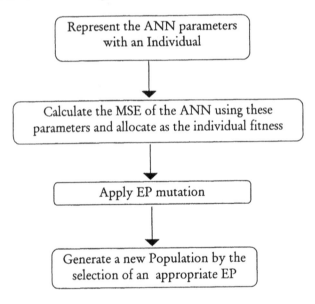

Figure 3.1 An evolution cycle of connection weights with EP

The *batch* training mode is adopted. In the batch training mode, weights are changed only after patterns have been presented to the EANN. This is different from most sequential training algorithms, like sequential BP, where weights are updated after each training pattern is presented to the network. The batch training mode is particularly suitable for performing the training in parallel.

3.5   REPRESENTATION OF CONNECTION WEIGHTS

Depending upon what evolution algorithms and what hardware platforms are used, the representation of weights can be selected. When using GA, the most convenient representation is binary, since GA uses binary representation (chromosomes) of the problem parameters and binary operators for combination. When using EP, a real number representation is more convenient since individuals are represented in EP as real numbers. The range of each parameter depends on the problem complexity and the required resolution of the parameters.

3.6   THE EVOLUTION OF AN EANN ARCHITECTURE

Section 3.4 assumed that an EANN architecture is predefined and fixed during the evolution of connection weights. But an EANN architecture has a significant impact on EANN information processing capabilities. An EANN architecture still has to be designed by experienced experts through trial and error. There is no report about a systematic way to design an optimal (near optimal) architecture for a particular task. There are two general algorithms, constructive and destructive. The first one starts with a minimal network (a network with a minimal number of hidden units and connections) and adds new nodes and connections if necessary during training. The second one does the opposite, i.e. starts with the maximal network and deletes unnecessary nodes and connections during training.

   The optimal design of an EANN architecture can be viewed as searching for an architecture that performs best on a specified task according to some criteria. There are several characteristics with such a surface that make the EP-based evolutionary approach a better candidate for searching than the heuristic approach, such as the aforementioned constructive/destructive algorithms. These characteristics include the fact that the surface is:

- *infinitely large* since the number of possible nodes and connections is unbounded;
- *non-differentiable* since changes in the number of nodes or connections is discrete and can have a discontinuous effect on EANN performance;
- *complex* and *noisy* since the mapping from an EANN architecture to EANN performance after training is indirect, and strongly dependent on initial conditions; and
- *multi-modal* since EANNs with quite different architectures can have very similar capabilities.

   The evolutionary design of architectures decides the proper representation of architectures. This representation can be in the form of a binary representation or a real one. For EP, a real representation is used. A general algorithm for the evolutionary design of architectures, in the form of the evolution of EANN connectivity, is presented in Figure 3.2 .

I.   Determine the parameters of the ANN including the Architecture into a set of Individuals.

II. Train each EPANN with the architecture by a predefined and fixed learning rate (but some parameters of the learning rule may be adaptable and learned during training), starting from different sets of random initial values of connection weights and, if any, training rule parameters.

III. Calculate the fitness of each individual on the basis of the above training results, e.g. on the basis of the smallest total MSE of training, the shortest training time or the architecture complexity (fewest nodes and connections and the like), etc.

IV. Reproduce new individuals in the current generation with probability according to their fitness.

V.   Apply the evolutionary operator to produce the new generation.

Figure 3.2   A typical algorithm for an EP-ANN

A key issue here is to decide how much information about an architecture is required for the representation. This includes the number of layers and the number of neurons in each layer. The more architecture parameters required in EA individuals, the greater computational cost. There is a trade-off between these two factors because the combination differs for different classes of problems.

3.7   THE EVOLUTION OF EANN TRAINING ALGORITHMS

Various types of ANNs needs different training algorithms. For example, EAs are suitable for training EANNs with feedback connections and deep feedforward EANNs (EANNs with many hidden layers), while BP is good at training shallow ones. Even after selecting a training algorithm, there are still algorithm parameters, such as the learning rate and momentum in BP algorithms, that need to be specified. The optimisation of training algorithms and their parameters for an EANN is usually very hard because little prior knowledge about an EANN architecture and the training task is available in practice.

Some work has been carried out on adaptively adjusting BP algorithm parameters, such as the learning rate and momentum, through the heuristic or evolutionary approach. However, the more fundamental issue of optimising the training algorithms, such as the weight-updating rule, has only been addressed by a limited number of researchers. It is very appealing to develop an automatic method to optimise training algorithms for an EANN. The evolution of the training ability of mankind from relatively weak to very powerful suggests the potential benefit of introducing an evolutionary process into EANN training. Figure 3.3 shows a typical algorithm for evolving training algorithms.

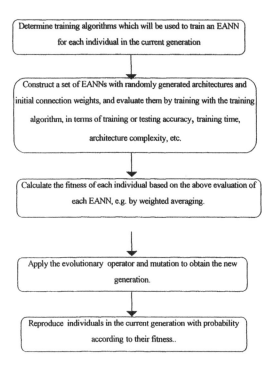

Figure 3.3 An algorithm for an EANN

Apart from offering an approach to optimising training algorithms, the evolution of training algorithms is also important in modelling the relationship between training, evolution and modelling the creative process. Newly evolved training algorithms have the potential to deal with a complex and changing environment. The fitness evaluation of each training algorithm is very noisy because randomness is introduced into the evaluation by not only the initial connection weights but also the architectures. Even if a particular architecture is predefined and fixed during evolution, noise still exists owing to the random initial connection weights.

3.8   THE EANN FRAMEWORK

Three evolution levels have been overviewed. Different levels of evolution react to different environments and use different time scales. The evolution of connection weights is at the lowest level and on the fastest time scale.

From an engineering view point, the decision on the level of evolution depends on the kind of available prior knowledge. If there is more prior knowledge about EANN architectures than about the training algorithms, then it is better to put the evolution of architectures at the highest level because such knowledge can be put into the architecture representation to reduce the search space. The lower level evolution of training algorithms can be more biased towards this type of architectures. On the other hand, the evolution of training algorithms should be at the highest level if there is more prior knowledge or if there is a special interest in certain types of training algorithms.

Figure 3.4 shows the amount of dominance at each level in an environment. For example, the evolution of architectures is decided solely by the task to be accomplished by the EANN. It can be viewed as a hierarchical adaptive system with three levels. At the highest level, architectures evolve at the slowest time scale decided by the task. For each architecture, there is a lower level evolution of training algorithms which proceeds at a faster time scale decided by the task as well as the architecture. As a result, the learning rule evolved is optimised towards the architecture. For each learning rule, there is an even lower level evolution of connection which proceeds at the fastest time scale in an environment decided by the task, architecture and learning rule.

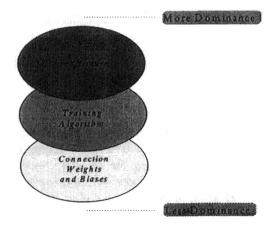

Figure 3.4   The level of evolution

Genetic algorithms, simulated annealing, gradient descent searches, evolution strategies and evolutionary programming can all be considered as types of evolutionary search procedures. The framework described in Figure 3.4 could be viewed as a hierarchical model of a general adaptive system for search procedures.

There are a few points worth mentioning about the general framework described by Figure 3.4. First, although the word *optimal* is normally used, it is very hard in practice to obtain an exact global optimum in a vast and complex space like those considered in this book. Fortunately, it is often the case in real-world applications that a good approximate solution (near optimal) is enough, not necessarily an exact global optimum. The criterion of 'good enough' varies from problem to problem. The evolutionary process is actually trying to find a near optimal solution, instead of an exact one.

Secondly, global search procedures like GAs are usually computationally expensive. That is why we do not always use GAs at all levels of evolution. It is, however, beneficial to introduce certain kinds of global searches at a particular level of an EANN. Especially when there is little prior knowledge available at that level and trial-and-error or other heuristic methods are very inefficient. As the power of parallel computers increases rapidly, the simulation of large EANNs becomes feasible. Such simulation will not only offer a better opportunity to discover novel EANN architectures and training algorithms, but also offer a way to model the creative process of EANN adaptation to a dynamic environment.

3.9 OO GA-ANN HYBRIDISATION

All the methods mentioned in the previous section can be employed to develop the GA-ANN system. The existing GA and ANN objects have been combined to form the new hybrid system. OO Development paradigm employed to construct the complete GA-ANN system including previously explained OO Analysis, OO Design, OO Implementation and OO Testing. New membership functions have been designed and existing classes are used. The system dynamic model is shown in Figure 3.5.

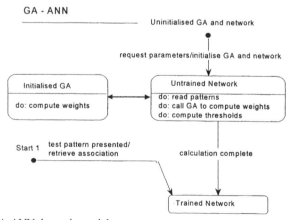

Figure 3.5 The GA-ANN dynamic model

After analysing the system, which includes identifying the object interactions, the adaptation of classes in the new environment is performed. This task also includes composing ANN parameters, including weights and biases, and decoding them into chromosomes. The number of ANN parameters is calculated using Equations (1) and (2) to form the chromosomes:

$$n_{free} = \left[(n_{in} \times n_{hid}) + (n_{hid} \times n_{out}) + n_{hid} + n_{out} \right] \tag{1}$$

$$S_{free} = n_{free} \times n_{byte} \tag{2}$$

where

n_{in} = number of nodes in the input layer,

n_{hid} = number of nodes in the hidden layer,

n_{out} = number of nodes in the output layer,

n_{byte} = number of bytes in each parameter,

n_{free} = number of free parameters,

S_{free} = size of free parameters in bytes.

Figure 3.6 shows the coding/decoding configuration of ANN parameters into chromosomes.

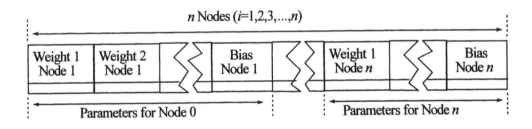

Figure 3.6 Coding/decoding for a GA-ANN

Other parameters such as architecture and training algorithms can be added to the chromosomes as an extension. The interaction between ANN and GA objects is performed by message passing. Both ANN and GA instances are created at the beginning of the optimisation procedure and last until the end. The GA object makes calls to ANN objects and passes messages to the fitness function.

3.10   OO EP-ANN HYBRIDISATION

EP and ANNs can be combined to create a more powerful intelligent system. The EP is used here to optimise the ANN parameters including weights, biases and architecture. When using EP to optimise ANNs modifiable, the error function of ANNs is used as the fitness function for EP. Both EP and ANN objects have been developed and described in detail in the previous chapters. The EP-ANN block diagram, Figure 3.7, shows the general interactions between these objects.

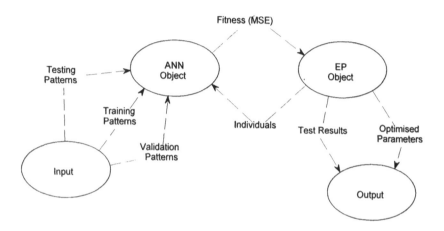

Figure 3.7   EP-ANN object interactions

As the EP is searching for a maximum fitness, this error is deducted from a large value in order to assign higher fitness to a set of ANN parameters with a lower MSE errors.   The ANN parameters are coded into individuals.   The individuals are presented in real number formats. In terms of computer implementation, they are coded into type Double.   The object Individual from EP object interacts with the object ANN by calling it in order to obtain the fitness. The EP standard mutation is performed by Equation (3) [2]:

$$\gamma_i = \gamma_i + \eta_i \tag{3}$$

where

γ_i = the ANN parameter (weight, bias, etc.)

$\eta_i \approx N(0,\sigma)$, $0<\sigma<1$,

$N(0,\sigma)$= normal distributed random variable, with mean 0 and variance σ.

Self-adaptive mutation could be introduced and the mutation parameters are represented by Equation (4):

$$\sigma_{i,j} = \sigma_{i,j}\exp[\tau'N(0,1) + \tau N_j(0,1)]$$

$$\tau = (\sqrt{2\sqrt{P}})^{-1}$$

$$\tau' = (\sqrt{2P})^{-1} \tag{4}$$

P = the ANN free parameters

The number of generations and the size of the population all depend on the problem complexity. There are two ways to utilise EP to optimise the ANN parameters. The first is to use EP to provide initial weight values for the ANN, then subsequently use another training method, backpropagation and a conjugate gradient method to complete the training phase. In this case, by combining EP and hill-climbing, a better compromise can be found to the trade-off of a global versus a local search. Therefore, the first solution for hybrid EP-ANN is a two-phase approach in which EP takes the place of the first phase of the search, providing the potential for non-convex optimisation. The second phase of this optimisation is subsequently applied to rapidly generate a precise solution under the assumption that the evolutionary search has generated a solution near the global optimum. The other method which is employed to create an EP-ANN hybrid system is only dependent on EP optimisation. In this case no other ANN training method is used and the ANN is solely trained by EP. The former method uses less computational processing time than the latter. However, there are strong beliefs that the latter, which demands higher computational power, gives a better solution. This means that a set of more optimised ANN parameters could be produced. Parallel computation is one solution to overcome this problem. These facilities might not be available for every developer but, after all, training is a one-off procedure. Once training is completed, the ANN can be easily executed to just propagate the data on normal PCs or any other available simple machines.

3.11 CONCLUSIONS

The object-oriented analysis, design and implementation of EANNs have been discussed. A number of issues were mentioned. The main question is: Are EANNs more efficient than gradient methods? A number of tests have been performed to investigate this. The OO methodology helped immensely to develop the tests and applications in a much shorter time by creating a very productive framework. The developed OO models of all algorithms allowed flexibility to upgrade and maintain the software constantly and form different configurations. If powerful computer facilities are available, then the EANN is generally the preferred method for creating ANNs. A trained intelligent system should be able to deal with new conditions and situations. A good system should be able to generalise its knowledge and expand it to unseen situations. Of course, this is far from what is achieved in today's world; however, research is being carried out to get us there in the future. The choice of training method is largely dependent on the problem characteristic. Different classes of problems have different attributes which should be examined to find the best method for optimisation. EANNs have generally longer training procedures, but in the applications shown later it is proven that they give better results. They are faster on small systems (up to 1400 parameters with our hardware systems) and create the best ANN performance. Once again it should be emphasised that results are application-dependent. In all the algorithms, one should make sure that enough information is represented to describe the problem.

3.12 REFERENCES

[1] T Bäck, U Hammel and H-P Schwefel, 'Evolutionary computation: comments on the history and current state', *IEEE Transactions on Evolutionary Computation*, **1**(1), April 1997, 3-17.

[2] P G Harrald and M Kamstra, 'Evolving artificial neural networks to combine financial forecasts', *IEEE Transactions on Evolutionary Computation*, **1**(1), April 1997, 40-52.

4

An Evolutionary Programming Approach to Reactive Power Planning

This chapter presents the application of evolutionary programming (EP) to reactive power planning (RPP). RPP is a non-smooth and non-differentiable optimisation problem for a multi-objective function. Several techniques to make EP practicable have been developed. The proposed approach is demonstrated with a modified IEEE 30-bus system and a practical UK power system. The comprehensive simulation results show that EP is a suitable method to solve the RPP problem. A conventional optimisation method and genetic algorithms are used as the comparison method. The comparison shows that EP gives the best solution for the RPP problem under the considered situations.

4.1   INTRODUCTION

Reactive power planning (RPP) is one of the most complex problems of power systems because it requires the simultaneous minimisation of two objective functions. The first objective deals with the minimisation of real power losses in reducing operating costs and improving the voltage profile. The second objective minimises the allocation cost of additional reactive power sources. RPP is a non-linear optimisation problem for a large-scale system with a lot of uncertainties. During the last decade there has been a growing concern with RPP problems [1-8]. Conventional calculus-based optimisation algorithms have been used in RPP for years [1-4]. Most conventional optimisation methods are based on successive linearisations and use the first and second derivatives of objective function and its constraint equations as the search directions. Because the formulae of RPP problems are hyperquadric functions, such linear and quadratic treatments induce many local minima. Furthermore, conventional optimisation methods cannot deal with the non-differentiable factor in the reactive power source installation function in RPP. Conventional optimisation methods can only lead to a local minimum and sometimes result in divergence in solving RPP problems. Recently, new methods based on artificial intelligence have been used in RPP or optimal reactive power control. Abdul-Rahman et al. [5] present an artificial neural network (ANN) enhanced by fuzzy sets to determine the membership of VAR control variables to solve load uncertainties, and an expert system to refine the solution with minimum adjustments of control variables. Jwo et al. [6] put forward a hybrid expert-system/simulated-annealing in RPP to solve local minimum problems. Genetic algorithms (GAs) are given in [7] and [8] for the global optimal solutions of reactive power optimisation problems. However, the ANN trained by linear programming [5] may still have the same problem of being stuck in a local minimum. With only one solution compared with another one to get the new solution in its iteration, simulated annealing (SA) would be more likely to converge either prematurely or keep searching without a direction. Mutalik et al. [9] have concluded, from their tested cases, that the GA performs consistently better than SA. The expert systems based on the analysis of sensitivities [5,6] are in the gradient directions to local minima.

This chapter proposes an application of evolutionary programming (EP) to RPP. EPs and GAs belong to evolutionary algorithms (EAs), which are search algorithms based on the simulated evolutionary process of natural selection and natural genetics [10-14]. EAs are randomised search algorithms which, however, do not necessarily mean directionless random walks. EAs are different from other optimisation methods because of the following features:

1. EAs search from a population of points, not a single point. The population can move over hills and across valleys. EAs can therefore discover a globally or near globally optimal point. Because the computation for each individual in the population is independent of others, EAs have an inherent parallel computation ability.
2. EAs use payoff (fitness or objective functions) information directly for the search direction, not derivatives or other auxiliary knowledge. EAs therefore can deal with non-smooth, non-continuous and non-differentiable functions that are the real-life optimisation problems. This property also relieves EAs of approximate assumptions for a many practical optimisation problems, which are quite often required in traditional optimisation methods.
3. EAs use probabilistic transition rules, not deterministic rules, to select generations, so they are a kind of stochastic optimisation algorithm which can search a complicated and uncertain area to find the global optimum. EAs are more flexible and robust than conventional methods.

These features make EAs robust and parallel algorithms to search adaptively for the globally optimal point. EAs offer new tools for the optimisation of complex system problems. EP is different from GAs in two ways: (i) EP uses the control parameters, not their codings; and (ii) the generation selection procedure of EP is mutation and competition, not reproduction, mutation and crossover. GAs emphasise models of genetic operators, while EP emphasises mutational transformations that maintain behavioural linkage. It has been indicated in [11] that EP outperforms GAs. The encoding and decoding of each solution and the operations of crossover and mutation on binary-coded variables of GAs use a lot of computing time. The new generation from GAs after mutation and crossover may lose advantages that is, it has a worse off solution as compared to that obtained in the last generation, while by competition in the combined old generation and mutated old generation, EP successfully takes such advantages.

The theory of EP has been well established, but some practical problems need to be solved to make EP practicable. In this chapter, some techniques have been developed to solve RPP and other practical problems:
1. Adaptive mutation scales are introduced to guarantee the global optimum and produce a smooth convergence.
2. Relative fitness values are used to deal with practical problems, of which the value of one individual does not differ much from the others.
3. Population size and competition size are carefully studied.

These techniques are essential in practical search problems.

A modified IEEE 30-bus system is used in this chapter as the simulation system. The results of EP are compared with those from genetic algorithms and a conventional method, Broyden-Fletcher-Goldfarb-Shanno method (BFGS) [15].

4.2   PROBLEM FORMULATION

List of symbols

N_1 number of load level durations

N_E number of branches

N_c number of possible reactive power source installation buses

N_i number of buses adjacent to bus i, including bus i

N_{PQ} number of PQ buses, which are load buses with constant P and Q injections

N_g number of generator buses

N_T numbers of tap-setting transformer branches

N_B total number of buses

N_{B-1} total number of buses, excluding slack bus

h per-unit energy cost (£/puWh, with $S_B = 100$ MVA)

d_1 duration of load level (hour)

g_k conductance of branch k (pu)

V_i voltage magnitude at bus i (pu)

θ_{ij} voltage angle difference between bus i and bus j (rad)

e_i fixed reactive power source installation cost at bus i (£)

C_{ci} per-unit reactive power source purchase cost at bus i (£/puVAR, $S_B = 100$

                    MVA)

Q_{ci} reactive power source installation at bus i (pu)

P_i, Q_i real and reactive powers injected into network at bus i (pu)

G_{ij}, B_{ij} mutual conductance and susceptance between bus i and bus j (pu)

G_{ii}, B_{ii} self-conductance and susceptance of bus i (pu)

Q_{gi} reactive power generation at bus i (pu)

T_k tap-setting of transformer branch k (pu)

N_{VPQlim} number of PQ buses at which the voltage violates the limits

N_{Qglim} number of buses at which the reactive power generation violates the limits

The objective function in RPP problems comprises two terms. The first term represents the total cost of energy loss as follows:

$$W_c = h \sum_{l \in N_l} d_l P_{loss}^l = h \sum_{l \in N_l} d_l \left(\sum_{\substack{k \in N_E \\ k=(i,j)}} g_k (V_i^2 + V_j^2 - 2V_i V_j \cos\theta_{ij}) \right)^l \quad (1)$$

where P_{loss}^l is the network real power loss during the period of load level l. The second term represents the cost of reactive power source installation which has two components, a fixed installation cost and a purchase cost:

$$I_c = \sum_{i \in N_c} (e_i + C_{ci}|Q_{ci}|) \quad (2)$$

where Q_{ci} can be either positive or negative, depending on whether the installation is capacitive or reactive. Therefore absolute values are used to compute the cost.

The above two objective functions are put into one equation to obtain a comprehensive one, which could be easily adjusted by changing the parameters in both W_c and I_c according to practical problems. The objective function can therefore be expressed as follows:

$$\min f_c = W_c + I_c$$

$$\text{s.t.} \quad 0 = P_i - V_i \sum_{j \in N_i} V_j (G_{ij}\cos\theta_{ij} + B_{ij}\sin\theta_{ij}) \qquad i \in N_{B-1}$$

$$0 = Q_i - V_i \sum_{j \in N_i} V_j (G_{ij}\sin\theta_{ij} - B_{ij}\cos\theta_{ij}) \qquad i \in N_{PQ}$$

$$Q_{ci}^{\min} \leq Q_{ci} \leq Q_{ci}^{\max} \qquad i \in N_c \tag{3}$$

$$Q_{gi}^{\min} \leq Q_{gi} \leq Q_{gi}^{\max} \qquad i \in N_g$$

$$T_k^{\min} \leq T_k \leq T_k^{\max} \qquad k \in N_T$$

$$V_i^{\min} \leq V_i \leq V_i^{\max} \qquad i \in N_B$$

where the power flow equations are used as equality constraints; and the reactive power source installation restrictions, the reactive power generation restrictions, the transformer tap-setting restrictions and the bus voltage restrictions are used as inequality constraints. The transformer tap-setting T, generator bus voltages V_g and reactive power source installations Q_c are control variables and so are self-restricted. The load bus voltages V_{load} and reactive power generations Q_g are state variables, which are restricted by adding them as the quadratic penalty terms to the objective function to form a penalty function. Equation (3) is therefore changed to the following generalised objective function:

$$\min F_c = f_c + \sum_{i \in N_{VPQlim}} \lambda_{Vi}(V_i - V_i^{\lim})^2 + \sum_{i \in N_{Qglim}} \lambda_{Qgi}(Q_{gi} - Q_{gi}^{\lim})^2$$

$$\text{s.t.} \quad 0 = P_i - V_i \sum_{j \in N_i} V_j (G_{ij}\cos\theta_{ij} + B_{ij}\sin\theta_{ij}) \qquad i \in N_{B-1} \tag{4}$$

$$0 = Q_i - V_i \sum_{j \in N_i} V_j (G_{ij}\sin\theta_{ij} - B_{ij}\cos\theta_{ij}) \qquad i \in N_{PQ}$$

where λ_{Vi} and λ_{Qgi} are the penalty factors which can be increased in the optimisation procedure; $V_i^{\lim}$ and $Q_{gi}^{\lim}$ are defined in the following equations:

$$V_i^{\lim} = \begin{cases} V_i^{\min} & \text{if} \quad V_i < V_i^{\min} \\ V_i^{\max} & \text{if} \quad V_i > V_i^{\max} \end{cases}$$

$$Q_{gi}^{\lim} = \begin{cases} Q_{gi}^{\min} & \text{if} \quad Q_{gi} < Q_{gi}^{\min} \\ Q_{gi}^{\max} & \text{if} \quad Q_{gi} > Q_{gi}^{\max} \end{cases} \tag{5}$$

It can be seen that the generalised objective function F_c is a non-linear and non-continuous function. The factor e_i in I_c is non-differentiable. Gradient-based conventional methods are not good enough to solve this problem.

4.3   EVOLUTIONARY PROGRAMMING (EP)

EP is different from conventional optimisation methods. It does not need to differentiate cost function and constraints. It uses probability transition rules to select generations. Each individual competes with some other individuals in a combined population of the old generation and the mutated old generation. The competition results are valued using a probabilistic rule. The individuals for the next generation are selected from the winners and constitute the same number as in the old generation. The procedure of EP for RPP is briefly listed as follows:

4.3.1 Initialisation

The initial control variable population is selected by randomly selecting $p_i=[V_g^i, Q_c^i, T^i]$, $i=1, 2, ..., m$, where m is the population size, from the sets of uniform distributions ranging over $[V^{min}, V^{max}]$, $[Q_c^{min}, Q_c^{max}]$ and $[T^{min}, T^{max}]$. The fitness value f_i of each p_i is obtained by running the PQ decoupled power flow.

4.3.2 Statistics

The values of the maximum fitness, minimum fitness, sum of fitnesses and average fitness of this generation are calculated as follows:

$$f_{max} = \{f_i \,|\, f_i \geq f_j \;\; \forall f_j, j = 1, ..., m\}$$

$$f_{min} = \{f_i \,|\, f_i \leq f_j \;\; \forall f_j, j = 1, ..., m\}$$

$$f_\Sigma = \sum_{i=1}^{m} f_i \tag{6}$$

$$f_{avg} = \frac{f_\Sigma}{m}$$

4.3.3 Mutation

Each p_i is mutated and assigned to p_{i+m} in accordance with the following equation:

$$p_{i+m,j} = p_{i,j} + N\left(0, \beta(x_{jmax} - x_{jmin})\frac{f_i}{f_{max}}\right) \quad j = 1,2,...,n \tag{7}$$

where $p_{i,j}$ denotes the jth element of the ith individual; $N(\mu, \sigma^2)$ represents a Gaussian random variable with mean μ and variance σ^2; f_{max} is the maximum fitness value of the old generation which is obtained in Section 4.3.2; x_{jmax} and x_{jmin} are the maximum and minimum limits of the jth element; and β is the mutation scale which is given as $0 < \beta \leq 1$. If any $p_{i+m,j}$, $j=1, 2, ..., n$, where n is the number of control variables, exceeds its limit, $p_{i+m,j}$ will be given the limit value. The corresponding fitness value f_{i+m} is obtained by running the power flow with p_{i+m}. A combined population is formed with the old generation and the mutated old generation.

4.3.4 Competition

Each individual p_i in the combined population has to compete with some other individuals to get its chance to be transcribed to the next generation. A weight value w_i is assigned to the individual according to the competition as follows:

$$w_i = \sum_{t=1}^{q} \begin{cases} 1 & \text{if } u_1 < \dfrac{f_t}{f_t + f_i} \\ 0 & \text{otherwise} \end{cases} \tag{8}$$

where q is the number of competitors; f_t is the fitness value of the tth randomly selected competitor in the combined population; f_i is the fitness value of p_i; u_1 is randomly selected from a uniform distribution set, $U(0,1)$. When all individuals p_i, $i=1, 2, ..., 2m$, get their competition weights, they will be ranked in descending order of their corresponding value w_i. The first m individuals are transcribed along with their corresponding fitness values f_i to be the basis of the next generation. The values of maximum, minimum and average fitness and sum of fitnesses of this generation are then calculated in the statistics process as in Section 4.3.2.

4.3.5 Inner Loop Convergence Criterion

The convergence is achieved when either the maximum fitness value converges to the minimum fitness value or the generations reach the maximum generation number. If the condition is met, the process will go to the next step; otherwise, the process will go back to the **Inner Loop Start**.

4.3.6 Outer Loop Convergence Criterion

The convergence is achieved when either all state variables, voltage magnitudes of load buses and reactive power generations, are within their limits or the outer loops reach the maximum number. If the condition is met, the program will stop. If one or more state variables violate their limits, the penalty factors of these variables will increase, and the process will go back to the **Outer Loop Start**.

To make EP practicable, the following four techniques have been developed [16]:

1. **Adaptive mutation scale**: In general, EP mutation probability is fixed throughout the whole search process. However, in practical applications, a

small fixed mutation probability can only result in a premature convergence, while a search with a large fixed mutation probability will not converge. To solve the problem, an adaptive mutation scale is given to change the mutation probability:

$$\beta(k+1) = \begin{cases} \beta(k) - \beta_{step} & \text{if } f_{min}(k) \text{ unchanged} \\ \beta(k) & \text{if } f_{min}(k) \text{ decreased} \\ \beta_{final} & \text{if } \beta(k) - \beta_{step} < \beta_{final} \end{cases}$$

$$\beta(0) = \beta_{init}$$

(9)

where k is the generation number; β_{init}, β_{final} and β_{step} are fixed numbers. β_{init} would be around 1 and β_{final} would be 0.005. β_{step} would be 0.001 to 0.01, depending on the maximum generation number. The mutation scale will decrease as the process goes on. The decreasing speed of the mutation scale depends on the fitness value; that is, the lower the fitness value, the faster the mutation scale decreases. Such an adaptive mutation scale not only prevents the premature convergence but also produces a smooth convergence.

2. **Relative fitness values**: In practical problems, the fitness value of one individual does not differ significantly from that of the others. Especially in the RPP problem, the difference between the minimum point and the original operating point is small. In deterministic transition rules, there may be no problem arising from this situation. However, in probabilistic transition rules, such a small difference will sink into oblivion because of added uncertainties, e.g. u_1 in EP. To deal with the problem, the program trims the fitness and the maximum fitness values that are used in the mutation and competition procedure. The method is explained as follows:

$$f_{proci} = f_i - \varepsilon f_{min} \qquad i = 1, 2, \ldots, m$$

$$f_{procmax} = f_{max} - \varepsilon f_{min}$$

(10)

where $0.95 \leq \varepsilon < 1$, so f_{proci} and $f_{procmax}$ will always be larger than 0. Only the relative fitness values are used in the process of mutation and competition. The relative values are quite distinct among the fitness values, so the better individuals become more competitive. It is the only way for EP to be practicable in real-life systems.

3. **Adaptive population size**: To ensure the global search area, the population size should increase with the increasing control variable number. According to our study experience, when the control variable number is very small, say, less than 10, keeping enough individuals in a population, say, 20 individuals, will not use too much computation time, because the generations will decrease with the decreasing variable number. However, when the control

variables increase, the population size does not need to increase in a direct ratio with the increase of variable number. The increase in population size will stop, for the sake of both computation speed and memory, when the size reaches a maximum number, which is given in this paper as 100. The relationship between the population size and the control variable number could be represented with the following equation:

$$Popsize = 20 \times Int\left(5 - 4e^{-k/40}\right) \qquad (11)$$

where k is the control variable number and Int means that only the integer of the value will be taken. The population size will be 20 for the control variables of 1-11, 40 for 12-27, 60 for 28-55 and 80 for more than 55.

4. **Competition size**: Since the competition size has very little influence on the computation speed and has no influence on storage, it could be kept high to obtain more information about the generation. However, a very high value of the competition size would result in a premature convergence. The competition size is given as 15 for a population size of 20, 20 for 40 and 25 for the larger population size. It is relatively large for a small population size in order to speed up the process and small for a large population size to avoid the premature convergence.

These techniques are essential in practical search problems. Without these techniques, EP cannot be applied to real-life problems. The population size in (11) and the competition size have been obtained by trial and error. Since the parameters for the four developed techniques depend only on the generation number and the number of control variables, they would also be applicable to other systems.

4.4   NUMERICAL RESULTS

In this section, the IEEE 30-bus system [17] is used to show the effectiveness of the algorithm. The network consists of six generator-buses, 21 load-buses and 43 branches, of which four branches, (6,9), (6,10), (4,12) and (28,27), are under-load-tap-setting transformer branches. The system is shown in Figure 4.1. The branch parameters and loads are given in [17]. The possible reactive power source installation buses are Buses 6, 17, 18 and 27. The base power and parameters of costs are given in Table 4.1. Three cases are studied. Case 1 is of light loads whose loads, and initial real power generations, except for the generation at the slack bus, are the same as those in [17]. Case 2 is of heavy loads, whose loads and initial real power generations are twice those of Case 1. Case 3 has two level load periods, one light load period having the same loads as those in Case 1 and one

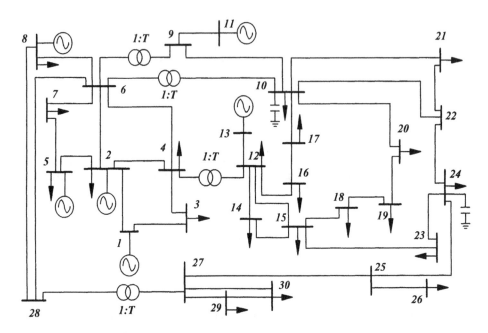

Figure 4.1 The modified IEEE 30-bus system

heavy load period having the same loads as those in Case 2. The one-year loss-of-energy cost is used to assess the possibility of installing the reactive power sources. The variable limits are given in Table 4.2. The total loads are:

Case 1: P_{load} = 2.834 pu, Q_{load} = 1.262 pu

Case 2: P_{load} = 5.668 pu, Q_{load} = 2.524 pu

4.4.1 Initial Power Flow Results

The initial generator bus voltages and transformer taps are set to 1.0 pu. The total initial generations and power losses are given in Table 4.3. The limit-violating quantities are given in Table 4.4. In Case 2, because of the heavy loads, all reactive power generations and almost all load bus voltages violate their limits.

Table 4.1 Base power and parameters of costs

S_B (MVA)	h (£/puWh)	e_i (£)	C_{ci} (£/puVAR)		d_l (hour)	
100	6000	1000	3000000	Case 1	Case 2	Case 3
				8760	8760	4380 for each level

Table 4.2 Variable limits (pu)

Bus	1	2	5	8	11	13
Q_g^{max}	0.596	0.480	0.6	0.53	0.15	0.155
Q_g^{min}	-0.298	-0.24	-0.3	-0.265	-0.075	-0.078
V^{max}	V^{min}	T^{max}	T^{min}	Q_c^{max}	Q_c^{min}	
1.05	0.95	1.1	0.9	0.36	-0.12	

Table 4.3 Initial generations and power losses (pu)

	P_g	Q_g	P_{loss}	Q_{loss}
Case 1	2.89388	0.98020	0.05988	-0.28180
Case 2	5.94588	3.26368	0.27788	0.73968

Table 4.4 Limit-violating variables (pu)

			Case 1			
Bus	26	29	30	Bus	8	
V_i	0.932	0.940	0.928	Q_{gi}	0.569	
			Case 2			
Bus	9	10	12	14	15	16	17
V_i	0.945	0.909	0.940	0.901	0.893	0.910	0.897
Bus	18	19	20	21	22	23	24
V_i	0.870	0.863	0.872	0.879	0.880	0.866	0.849
Bus	25	26	27	29	30		
V_i	0.855	0.811	0.880	0.829	0.799		

Table 4.4 continued

Bus	1	2	5	8	11	13
Q_{gi}	-0.402	0.496	0.952	1.497	0.281	0.439

4.4.2 EP Optimal Results

The optimal results are given in Table 4.5. The transformer taps are discrete variables with a change step of 0.025 pu. All the state variables are regulated back into their limits.

Table 4.5 Optimal control variables

Generator Bus Voltages (pu)						
Bus		1	2	5	8	11	13
Case 1		1.050	1.044	1.023	1.025	1.050	1.050
Case 2		1.050	1.022	0.973	0.959	1.050	1.050
Case 3	Light load	1.050	1.044	1.023	1.026	1.027	1.050
	Heavy load	1.050	1.022	0.973	0.959	1.045	1.048

Transformer Tap-settings (pu)					
Branch		(6,9)	(6,10)	(4,12)	(28,27)
Case 1		0.95	1.1	1.025	1.05
Case 2		1.05	1.1	1.1	1.1
Case 3	Light load	0.975	1.1	1.05	1.025
	Heavy load	1.05	1.075	1.1	1.1

Reactive Power Source Installations (pu)					
Bus		6	17	18	27
Case 1		0	0	0	0
Case 2		0.198	0.229	0.133	0.196
Case 3	Light load	0.150	0.077	0.068	0.077
	Heavy load	0.198	0.227	0.131	0.195

Power generations and Power Losses (pu)					
		P_g	Q_g	P_{loss}	Q_{loss}
Case 1		2.88559	0.92579	0.05159	-0.33626
Case 2		5.90110	2.20392	0.23311	0.43591
Case 3	Light load	2.88485	0.52662	0.05085	-0.36343
	Heavy load	5.90195	2.21541	0.23396	0.44257

Table 4.5 continued

Iteration and Computation Time (486/50 MHz)

Outer Loop		1	2	3	4	5	Time (min)
Generations (Iterations)	Case 1 85		108				4.2
	Case 2 91		112	134	97		8.5
	Case 3 97		129	106	141	93	20.21

Two sets of control variables are obtained for two different load level periods in Case 3. The PQ decoupled power flow has to run twice for two sets of control variables, so the computation time almost doubles for the same generation as in Case 1 or 2. In Case 3, because the installation cost is only counted in the heavy load period, which has more installation than the light load period for every bus, the reactive power sources used in the light load period do not induce any cost. Therefore, in the light load period, there are some reactive power source installations, which are in fact the reactive power generations from the existing reactive power sources installed for the heavy load period. The real power loss is lower than that in Case 1 because of these reactive power sources. The real power savings and annual cost-of-energy savings are given as follows:

Case 1:

$$P_{save}\% = \frac{P_{loss}^{init} - P_{loss}^{opt}}{P_{loss}^{init}} \times 100 = \frac{0.05988 - 0.05159}{0.05988} \times 100 = 13.84\%$$

$$W_c^{save} = hd_l(P_{loss}^{init} - P_{loss}^{opt})$$
$$= 6000 \times 8760 \times (0.05988 - 0.05159)$$
$$= \pounds435722.4$$

Case 2:

$$P_{save}\% = \frac{0.27788 - 0.23311}{0.27788} \times 100 = 16.11\%$$

$$W_c^{save} = 6000 \times 8760 \times (0.27788 - 0.23311) = \pounds2353111.2$$

Case 3:

Light load period: $P_{save}\% = \dfrac{0.05988 - 0.05085}{0.05988} \times 100 = 15.08\%$

Heavy load period: $P_{save}\% = \dfrac{0.27788 - 0.23396}{0.27788} \times 100 = 15.81\%$

$$W_c^{save} = 6000 \times 4380 \times [(0.05988 - 0.05085) + (0.27788 - 0.23396)]$$
$$= £1391526$$

4.4.3 Comparison with the BFGS Method

A non-linear programming algorithm, the BFGS method, is used as the comparison method. After the optimisation of BFGS, the real power losses are 0.05470, 0.23483, 0.05729 and 0.24070 per-unit in Cases 1 and 2 and the light load period and heavy load period of Case 3, respectively. The total reactive power installations are 0.265, 1.071, 0.567 and 0.959 per-unit in Cases 1 and 2 and the light load period and heavy load period of Case 3, respectively. The power savings with the BFGS method are 8.65% in Case 1, 15.50% in Case 2 and 4.33% in the light load period and 12.38% in the heavy load period of Case 3. The annual energy cost savings are £272260.8, £2263233.6 and £1045155.6 in Cases 1, 2 and 3, respectively. The computation times are 0.8, 1.5 and 3.7 min in the three cases, respectively. The annual cost-of-energy savings from EP are 160%, 104% and 133% of those from BFGS in the three cases respectively.

The total cost of energy losses and investments, which is the objective of RPP in Equation (4), are given as follows:

Case 1:   EP: $f_C = W_C + I_C = £2711570 + 0 = £2711570$

BFGS: $f_C = W_C + I_C = £2875032 + £799000 = £3674032$

Case 2:   EP: $f_C = W_C + I_C = £12252262 + £2272000 = £14524262$

BFGS: $f_C = W_C + I_C = £12342665 + £3217000 = £15559665$

Case 3:   EP: $f_C = W_C + I_C = £7484807 + £2257000 = £9741807$

BFGS: $f_C = W_C + I_C = £7831177 + £2881000 = £10712177$

In Case 3, each installation bus has two reactive power source installations for two load level periods. The larger installation is used to compute the installation cost I_C, while the smaller quantity can be obtained by regulation during the light load period. In both EP and BFGS, the installation at any bus in the heavy load period is larger than the installation in the light load period, so the installation cost of Case 3 is the cost for the heavy load period.

The total costs from EP are 74%, 93% and 91% of those from BFGS in the three cases, respectively. From the comparison, it can be seen that in all three cases, EP gives better results. Therefore, in this simulation, EP always goes to the global or near global optimum, while BFGS goes to a local minimum.

4.5   PRACTICAL RESULTS

The proposed approach has been applied to a real power system in the UK. Simulation results, compared with those obtained by using an improved genetic algorithm (GA) and a conventional gradient-based optimisation method, BFGS method, are presented to show that the present method is better for power system planning. Several cases simulating the real network situation of both a normal operation and an operation with line outages have been studied. For all these cases, EP is a much better method than the others.

This section shows the benefit obtained in using EP for RPP in a practical power system. The practical network consists of 40 buses and 56 branches. The buses comprise eight generator buses that are represented by PV buses and a slack bus and 18 load buses. There are 11 under-load-tap-changing transformers and five possible VAR source installation buses. There are 24 control variables in total. Though the system is not big enough to show that EP can solve the RPP problem for a real large power system, it at least shows that EP is capable of dealing with the problems of a real power system under real operation states. It is believed that for a large-scale system, the main difference is the increase in computation time and memory requirements.

The network parameters are given in Tables 4.A-1 and 4.A-2 of the Appendix. The following seven cases have been studied:

A uniform load profile for 12 months at 100% of the load level.
A uniform load profile for 12 months at 130% of the load level.
A stepped load profile with six months at 100% and six months at 130% of the load level.
A uniform load profile for 12 months at 100% of the load level, with the circuit outages of (11-12) and (11-38).
A uniform load profile for 12 months at 100% of the load level, with the circuit outage of (12-15).
A uniform load profile for 12 months at 130% of the load level, with the circuit outages of (11-12) and (11-38).
A uniform load profile for 12 months at 130% of the load level, with the circuit outage of (12-15).

The loads in Cases 1, 4 and 5 are
$$P_{load} = 41.1534 \qquad Q_{load} = 7.073$$

The loads in Cases 2, 6 and 7 are
$$P_{load} = 53.4994 \qquad Q_{load} = 9.1949$$

One year's loss-of-energy cost is employed to assess the possibility of installing the VAR sources. The variable limits and duration of load levels are given in Table 4.A-3 of the Appendix.

The initial generations and power losses and the variable violations are obtained as in Table 4.6. In Cases 4-7, with the same initial conditions as in Cases 1 and 2, more reactive power generations go out of their limits. In Case 6, 14 buses are running under lower voltage limits, which can be seen as serious security problems in the network. The initial conditions of Case 3 are the same as those for Cases 1 and 2.

Table 4.6 Initial power flow results

Generations and Power Losses			
	P_g	Q_g	P_{loss}	Q_{loss}
Case 1	41.4661	1.9938	0.3127	-5.0792
Case 2	54.0514	10.6546	0.5520	1.4599
Case 4	41.5897	4.4044	0.4363	-2.6685
Case 5	41.7271	4.8165	0.5734	-2.2565
Case 6	54.3195	15.3653	0.8200	6.1708
Case 7	54.5568	16.1692	1.0574	6.9743

Voltages Outside Limits							
Bus	1	2	3	4	5	7	8
Case 6 V_i	0.91	0.91	0.92	0.91	0.92	0.94	0.93
Bus	9	10	11	29	30	31	34
V_i	0.91	0.91	0.90	0.86	0.85	0.87	0.88

Reactive Power Generations Outside Limits			
Bus	24	39	40
Case 1			1.209
Case 2	1.540	3.432	2.531
Case 4		1.216	1.336
Case 5		1.422	1.396
Case 6	2.559	3.278	2.560
Case 7	3.163	4.974	3.028

Five buses, 6, 13, 26, 29 and 34, are selected as the possible VAR source installation buses. The generic range from -75 to +150 MVAR is used as the VAR source installation limit. The costs of energy loss and installation can be found in Table 4.A-3 of the Appendix. An improved genetic algorithm (GA) [18] and a conventional optimisation method, BFGS method, are used as the comparison methods. The optimal generator bus voltages, VAR source installations and generations and power losses are given in Tables 4.7 to 4.10. The transformer tap-settings are given in Table 4.11. In all the cases, the voltage and reactive power violations have been eliminated.

Table 4.12 gives the energy saved and cost comparisons between EP and BFGS, which is obtained from the following equation:

$$W_c^{save} = hd_1 (P_{loss}^{init} - P_{loss}^{opt})$$
$$F_c = I_c + W_c$$

Table 4.7 Optimal results (generator bus voltages)

Bus		18	20	22	24	39	40
Case 1	EP	1.04	1.07	1.06	1.07	1.06	1.05
	GA	1.04	1.07	1.07	1.06	1.06	1.05
	BFGS	1.01	1.03	1.03	1.02	1.02	1.00
Case 2	EP	1.08	1.04	1.05	1.06	1.04	1.01
	GA	1.08	1.03	1.05	1.06	1.04	1.02
	BFGS	1.09	0.94	1.03	1.01	1.00	1.02
Case 3	EP	1.09	1.02	1.03	1.06	1.04	1.02
	GA	1.05	1.03	1.06	1.06	1.04	1.04
	BFGS	1.08	1.06	1.04	1.00	0.98	0.98
Case 4	EP	1.05	1.06	1.04	1.06	1.06	1.06
	GA	1.05	1.07	1.06	1.06	1.06	1.05
	BFGS	1.05	1.01	1.04	1.02	1.01	1.00
Case 5	EP	1.05	1.05	1.06	1.05	1.06	1.05
	GA	1.05	1.07	1.08	1.05	1.05	1.05
	BFGS	1.06	1.04	1.05	1.00	1.04	1.03
Case 6	EP	1.08	1.07	1.09	1.05	1.04	1.04
	GA	1.09	1.07	1.06	1.06	1.04	1.04
	BFGS	1.07	1.03	1.03	1.02	1.01	1.03
Case 7	EP	1.09	1.04	1.09	1.04	1.05	1.06
	GA	1.09	1.06	1.07	1.04	1.05	1.05
	BFGS	1.06	0.99	0.96	0.98	1.02	1.01

Table 4.8 Optimal results (VAR source installations)

Bus		6	13	26	29	34
Case 1	EP	0	0	0	0	0
	GA	0	0	0	0	0
	BFGS	0.031	0.090	0.273	-0.361	-0.131
Case 2	EP	0	0.469	0.360	0	0
	GA	0.004	0.537	0.291	0.001	0.045
	BFGS	-0.330	0.215	1.490	0.480	-0.100
Case 3	EP	-0.134	0.343	0.482	-0.008	0.064
	GA	-0.744	0.661	1.500	0.207	0.689
	BFGS	-0.010	-0.025	1.510	0.014	0.345
Case 4	EP	0	0	0.011	0	0
	GA	0	0.023	0.035	0	0
	BFGS	-0.121	-0.570	0.343	0.074	-0.009

Table 4.8 continued

Case 5	EP	0	0	0	0	0
	GA	0	0	0	0	0
	BFGS	0.033	-0.289	-0.072	-0.186	-0.020
Case 6	EP	0	1.002	1.497	0	0.002
	GA	0	0.999	1.469	0	0
	BFGS	-0.291	0.272	1.500	-0.510	-0.752
Case 7	EP	0.009	1.224	1.500	0	0.417
	GA	0.022	0.885	1.464	0.005	0.753
	BFGS	-0.430	0.775	1.490	-0.314	1.041

Table 4.9 Optimal results (generations and power losses)

		P_g	Q_g	P_{loss}	Q_{loss}
Case 1	EP	41.4059	-0.6530	0.2743	-7.7385
	GA	41.4074	-0.6407	0.2749	-7.7234
	BFGS	41.4502	1.2184	0.2996	-6.0658
Case 2	EP	53.9653	6.2594	0.4839	-2.11305
	GA	53.9667	6.1231	0.4841	-2.1936
	BBFGS	54.0190	7.1217	0.5149	-0.3322
Case 3	EP	41.4020	-1.2180	0.2806	-7.5457
100% Base Load	GA	41.4312	-2.9131	0.2898	-7.673
	BFGS	41.4681	-0.5019	0.3021	-5.7626
Case 3	EP	53.8906	6.4781	0.4911	1.9715
130% Base Load	GA	53.8797	4.7339	0.5003	-2.1485
	BFGS	54.0183	7.8697	0.5361	0.3109
Case 4	EP	41.4302	1.3176	0.3764	-5.7730
	GA	41.4294	1.2481	0.3762	-5.7676
	BFGS	41.5275	3.0974	0.4149	-4.0604
Case 5	EP	41.6440	1.6891	0.5000	-5.3838
	GA	41.6471	1.8728	0.5032	-5.2005
	BFGS	41.6604	3.3103	0.5210	-4.5365
Case 6	EP	54.0575	7.0583	0.6475	0.3927
	GA	54.0582	7.0732	0.6486	0.4469
	BFGS	54.1485	12.9958	0.7470	4.1623
Case 7	EP	54.2644	7.2582	0.8549	1.3044
	GA	54.2669	7.3707	0.8573	1.4044
	BFGS	54.4380	12.0052	0.9787	5.3514

Table 4.10 Optimal results (iteration and computation time in minutes for 486/50 MHz PC)

	Case 1	Case 2	Case 3	Case 4	Case 5	Case 6	Case 7
EP	9.5	10.3	24.8	12.7	11.4	14.1	13.6

Table 4.10 continued

GA	16.2	19.9	41.3	21.9	20.0	25.2	24.3
BFGS	2.3	2.5	4.9	2.8	2.7	3.0	2.9

Table 4.11 Transformer tap-settings

Branch		(1-29)	(2-30)	(3-31)	(4-31)	(5-32)	(9-35)	(10-33)	(11-34)	(8-36)	(27-37)	(28-37)
Case 1	EP	1.1	1.05	1.05	1.05	1.075	1.025	0.95	1.025	1.00	0.975	0.975
	GA	1.1	1.05	1.0	1.0	1.025	1.025	0.9	1.05	1.0	0.975	0.975
	BFGS	1.05	0.95	1.025	1.05	1.05	1.0	1.0	1.05	1.025	0.975	1.0
Case 2	EP	1.05	1.05	1.025	1.025	1.05	1.0	0.975	1.075	1.025	1.05	1.075
	GA	1.1	1.1	1.025	1.075	1.0	0.975	1.05	1.075	1.0	1.1	1.075
	BFGS	1.05	1.0	1.0	1.0	1.0	0.925	0.975	0.95	0.925	1.1	0.95
Case 3	EP	1.1	1.05	1.1	0.975	1.025	0.95	1.025	1.05	0.95	1.0	1.0
	GA	1.1	1.075	1.05	1.05	0.975	1.025	1.025	1.05	1.025	1.0	1.1
	BFGS	1.05	1.0	0.95	1.075	1.05	1.05	0.95	0.95	1.025	1.025	1.0
Case 4	EP	1.025	1.025	1.075	1.025	1.025	0.975	1.025	1.05	1.075	1.05	1.025
	GA	1.05	1.0	1.025	1.025	1.025	1.05	0.975	1.05	1.075	1.05	1.025
	BFGS	0.95	1.075	1.0	1.025	0.975	1.025	1.0	1.025	1.0	1.05	1.0
Case 5	EP	1.075	1.075	1.075	1.05	1.05	0.925	1.0	1.05	0.975	0.95	0.95
	GA	1.0	1.025	1.075	1.05	0.975	0.925	1.025	1.05	1.05	1.025	1.025
	BFGS	1.05	1.0	1.0	1.025	1.025	1.05	1.0	1.0	0.975	1.075	1.0
Case 6	EP	1.1	1.05	1.025	0.95	1.025	1.025	0.925	1.075	1.0	1.1	1.05
	GA	1.1	1.0	1.025	1.0	1.025	0.9	0.975	1.025	1.05	1.05	1.075
	BFGS	1.025	0.975	1.05	1.025	1.025	1.0	1.025	0.975	1.05	1.05	0.975
Case 7	EP	1.1	1.05	0.975	1.025	1.025	1.05	1.025	1.0	1.025	1.05	1.1
	GA	1.075	1.05	1.0	1.0	1.025	1.0	1.0	1.0	1.025	1.075	1.075
	BFGS	0.95	1.0	0.975	1.0	1.0	1.0	1.0	0.95	0.975	0.975	1.0

Table 4.12 Comparison between EP, GA and BFGS

		W_C^{save} (£/year)	F_C (£)
Case 1	EP	2,018,304	14,419,208
	GA	1,986768	14,601,400
	BFGS	688,536	17,125,600
Case 2	EP	3,579,336	26,770,400
	GA	3,568,824	26,638,300
	BFGS	1,949,976	29,921,800
Case 3	EP	2,444,040	22,524,700
	GA	1,960,488	24,728,300
	BFGS	696,420	25,480,700
Case 4	EP	3,148,344	19,814,800
	GA	3,158,856	19,825,500
	BFGS	1,124,784	23,963,100

Table 4.12 continued

Case 5	EP	3,857,904	26,277,300
	GA	3,689,712	26,477,400
	BFGS	2,754,144	28,617,100
Case 6	EP	9,066,600	37,222,100
	GA	9,008,784	37,320,500
	BFGS	3,836,880	42,507,000
Case 7	EP	10,643,400	48,627,600
	GA	10,517,256	48,781,700
	BFGS	4,136,472	56,321,400

4.6   CONCLUSIONS

This is the first time that the application of EP to the RPP of power systems has been reported. RPP is an optimisation problem of a non-linear, non-smooth and non-continuous function. This type of less well-behaved function is met with in most engineering problems, so the devising, testing and refining of new techniques to find optimal solutions becomes more important nowadays with the advent of even more powerful computers. The proposed EP approach has been evaluated on an IEEE 30-bus power network. The simulations show that EP always leads to satisfactory results of multiobjective RPP in different cases, especially in non-continuous and non-smooth situations. The comparison shows that the proposed EP method is more powerful for global optimisation problems of the non-smooth, non-continuous functions. The only disadvantage of EP is that it takes much more computation time than the conventional method. However, with an inherent parallel computation ability and the advance of computer technology, it would not be so difficult to solve such a problem. The comprehensive simulation results show great potential for applications of EP in economical and secure power system operations, planning and the assessment of reliability.

In all 'real-life' simulation cases, both EP and a GA have obtained much better results than BFGS method. It can be concluded that both EP and a GA have reached global or near-global minimum points, while BFGS method has stuck at some local minimum points. The results of EP are a little better than those of a GA. However, both EP and a GA use much more computation time than BFGS. Between a GA and EP, the former uses almost twice the computation time as does the latter. Therefore, EP is better than other methods.

4.7   ACKNOWLEGMENTS

The author wishes to acknowledge the IEEE and IEE for granting permission to reproduce the material and results contained in [19] and [20] respectively.

4.8 REFERENCES

[1] A Kishore and E F Hill, 'Static optimization of reactive power sources by use of sensitivity parameters', *IEEE Transactions on Power Apparatus and Systems*, **PAS-90**, 1971, 1166-1173.

[2] S S Sachdeva and R Billinton, 'Optimum network VAR planning by nonlinear programming', *IEEE Transaction on Power Apparatus and Systems*, **PAS-92**, 1973, 1217-1225.

[3] R A Fernandes, F Lange, R C Burchett, H H Happ and K A Wirgau, 'Large scale reactive power planning', *IEEE Transactions on Power Apparatus and Systems*, **PAS-102**, 1983, 1083-1088.

[4] Y Y Hong, D I Sun, S Y Lin and C J Lin, 'Multi-year multi-case optimal VAR planning', *IEEE Transactions on Power Systems*, **PWRS-5**, 1990, 1294-1301.

[5] K H Abdul-Rahman, S M Shahidehpour and M Daneshdoost, 'AI approach to optimal VAR control with fuzzy reactive loads', *IEEE Transactions on Power Systems*, 1995, **PWRS-10**, 88-97.

[6] W S Jwo, C W Liu, C C Liu and Y T Hsiao, 'Hybrid expert system and simulated annealing approach to optimal reactive power planning', *IEE Proceedings - Generation, Transmission and Distribution*, **142**, 1995, 381-385.

[7] K S Swarup, M Yoshimi and Y Izui, 'Genetic algorithm approach to reactive power planning in power systems', *Proceedings of the 5th Annual Conference of Power & Energy Society IEE Japan*, 1994, 119-124.

[8] K Iba, 'Reactive power optimization by genetic algorithm', *IEEE Transactions on Power Systems*, **PWRS-9**, 1994, 685-692.

[9] P P Mutalik, L R Knight, J L Blanton and R L Wainwright, 'Solving combinatorial optimization problems using parallel simulated annealing and parallel genetic algorithms', *Proc. of the 1992 ACM/SIGAPP Symposium on Applied Computing*, USA, 1992, 1031-1038.

[10] L J Fogel, 'Autonomous automata', *Industrial Research*, **4**, 1962, 14-19.

[11] D B Fogel, *System Identification Through Simulated Evolution: A Machine Learning Approach to Modeling*, Ginn Press, MA, 1991.

[12] D B Fogel, 'An introduction to simulated evolutionary optimization', *IEEE Transactions on Neural Networks*, **5**, 1994, 3-14.

[13] P J Angeline, 'Evolution revolution: an introduction to the special track on genetic and evolutionary programming', *IEEE Expert, **10**, June 1995*, 6-10.

[14] D B Fogel, *Evolutionary Computation: Toward a New Philosophy of Machine Intelligence*, IEEE Press, Piscataway, NJ, 1995.

[15] M S Bazaraa, H D Sherali and C M Shetty, *Nonlinear Programming: Theory and Algorithms*, John Wiley & Sons, 1993, 318-328.

[16] J T Ma and L L Lai, 'Evolutionary programming approach to reactive power planning', *IEE Proceedings - Generation, Transmission and Distribution*, **143**, 1996, 365-370.

[17] O Alsac and B Stott, 'Optimal load flow with steady-state security', *IEEE Transactions on Power Apparatus and Systems*, **93**, 1974, 745-751.

[18] J T Ma and L L Lai, 'Improved genetic algorithm for reactive power planning', *12th Power Systems Computation Conference*, **1**, Dresden, Germany, August 1996, 499-505.

[19] L L Lai and J T Ma, 'Application of evolutionary programming to reactive power planning - comparison with non-linear programming approach', *IEEE Transactions on Power Systems*, **12**, No 1, Feb 1997, 198-206.

[20] L L Lai and J T Ma, 'Practical application of evolutionary computing to reactive power planning', *IEE Proceedings - Generation, Transmission and Distribution*. (accepted for publication).

4.9   APPENDIX

The network parameters are given in Tables 4.A-1 and 4.A-2. All the variables are per-unit quantities based on S_B = 100 MVA. The schematic diagram of the UK power network is shown in Figure 4.2.

Table 4.A-1 Bus parameters

Bus	T^*	P_g	Q_{min}	Q_{max}	P_l	Q_l	V
1	0	0	0	0	0	0	1.021
2	0	0	0	0	0	0	1.022
3	0	0	0	0	0	0	1.025
4	0	0	0	0	0	0	1.024
5	0	0	0	0	0	0	1.02
6	0	0	0	0	0	0	1.02
7	0	0	0	0	0	0	1.023
8	0	0	0	0	0	0	1.019
9	0	0	0	0	0	0	1.02
10	0	0	0	0	0	0	1.02
11	0	0	0	0	0	0	1.017
12	0	0	0	0	4.22	-0.638	1.01
13	0	0	0	0	0	0	1.01
14	0	0	0	0	2.004	0.484	1.011
15	0	0	0	1.20	4.619	0.965	1.009
16	0	-1.088	0	0	0	0	1.011
17	0	0	0	0	0	0	1.023
18	1	10.991	-3.60	6.60	0	0	1
19	0	0	0	0	0	0	1.02
20	1	2.078	-0.90	1.50	0	0	1
21	0	0	0	0	0	0	1.02
22	1	2.078	-0.90	1.50	0	0	1
23	0	-3.926	0	0	0	0	1.02
24	1	-0.386	-1.0	1.0	0	0	1.005
25	0	0	0	0	5.185	1.173	1
26	0	0	0	0	3.584	1.531	0.995
27	0	0	0	0	0	0	1.024
28	0	0	0	0	0	0	1.024
29	0	0	0	0	2.9482	0.637	1.075
30	0	0	0	0	1.3162	0.3912	1.014
31	0	0	0	0	1.5794	0.4694	1.013
32	0	0	0	0	1.4215	0.2235	1.053
33	0	0	0	0	0.4001	0.0559	1.075
34	0	0	0	70	3.6852	0.9164	1.033
35	0	0	0	0	0.8423	0.2235	1.053
36	0	0	0	0	0.3159	0.0671	1.042
37	0	0	0	0	2.0006	0	1.011
38	0	0	0	0	1.56	0.574	1.013
39	1	18.065	-1.30	1.0	0	0	1
40	2	8.179	-1.0	1.0	0	0	0.996

* 0 = PQ Bus, 1 = PV Bus, 2 = Slack Bus.

Table 4.A-2 Branch data

Bus I	Bus J	$R(\%)$	$X(\%)$	$B(\%)$	TAP
1	29	0.1418	8.3418	0	1.1
2	30	0.1473	7.9666	0	1.0
3	31	0.1628	8.4166	0	0.95
4	31	0.1614	8.3541	0	0.95
5	32	0.1701	8.3125	0	1.1
9	35	0.075	3.88	0	1.05
10	33	0.073	4.12	0	1.05
11	34	0.04	2.056	0	1.025
8	36	0.1458	8.0458	0	1.1
27	37	0.1859	8	0	0.95
28	37	0.1697	8.0875	0	0.95
29	30	0	11.721	0	1
31	32	0	3.4641	0	1
31	30	0	7.906	0	1
32	36	0	10.3801	0	1
37	36	0	2.2255	0	1
27	19	0.0944	0.759	12.4715	1
28	21	0.0943	0.7571	11.7106	1
1	2	0.1173	0.9338	28.7869	1
1	3	0.2284	1.8398	55.46	1
2	4	0.1106	0.9019	26.5625	1
3	5	0.118	0.9832	27.72	1
4	5	0.1173	0.9776	27.5683	1
5	9	0.0436	0.5956	24.7305	1
7	17	0.0641	0.5102	15.7178	1
8	17	0.0641	0.5102	15.7178	1
9	11	0.1122	1.5332	63.6572	1
5	10	0.0909	1.2123	50.2355	1
10	11	0.0645	0.8821	36.6269	1
11	12	0.0963	1.3157	54.6167	1
11	38	0.0728	0.9951	41.29	1
12	14	0.0483	0.6598	27.3966	1
12	38	0.0306	0.3964	76.04	1
12	15	0.0936	0.7448	22.9675	1
12	25	0.0779	1.0648	44.21	1
14	38	0.0246	0.3362	13.9605	1
15	16	0.0211	0.2885	11.9788	1
16	23	0.1964	1.5628	48.192	1
16	24	0.0523	0.713	75.47	1
17	19	0.035	2.4146	5.8978	1
17	1	0.035	2.4146	5.8978	1
17	23	0.2077	1.6536	50.98	1
25	26	0.0617	0.8432	35.01	1

Table 4.A-2 continued

26	40	0.052	0.711	29.5	1
1	7	0.409	3.4082	96.1112	1
1	8	0.409	3.4082	96.11	1
5	7	0.0918	0.7305	22.515	1
5	8	0.0918	0.7305	22.515	1
18	17	0.0115	1.055	0	1
39	38	0.0108	0.6375	0	1
20	19	0.205	8.05	0	1
22	21	0.205	8.05	0	1
13	12	0	3.3333	0	1
6	5	0	3.3333	0	1
12	38	0.0306	0.3964	76.04	1
16	24	0.0523	0.713	75.47	1

NB R, X and B are percentage quantities on a 100 MVA base.
The following points should be noted:

1. The voltage limits are changed to ±6% for supergrids of 275 kV and 400 kV and ±10% for 132 kV and terminal buses of equivalent generators.
2. Buses 6, 13, 26, 29 and 34 are selected as potential VAR source installation sites.
3. The generic range from -75 to +150 MVAR is used as the VAR source installation limits.
4. The 400 kV transmission line outages, lines (11-12) and (11-38) and line (12-15), have been considered.

Table 4.A-3 Case study parameters

Voltage and Tap-setting Limits					
400 and 275 kV Bus		132 kV and Below			
V max	V min	V max	V min	T max	T min
1.06	0.94	1.1	0.9	1.1	0.9
Duration of Load Level					
	Cases 1-2, 4-7		Case 3		
d_l (hour)	8760		4380 for each level		
Costs of Energy Loss and Installation					
h (£/kWh)		e_i (£)		C_{ci} (£/kVAR)	
0.06		1000		10	

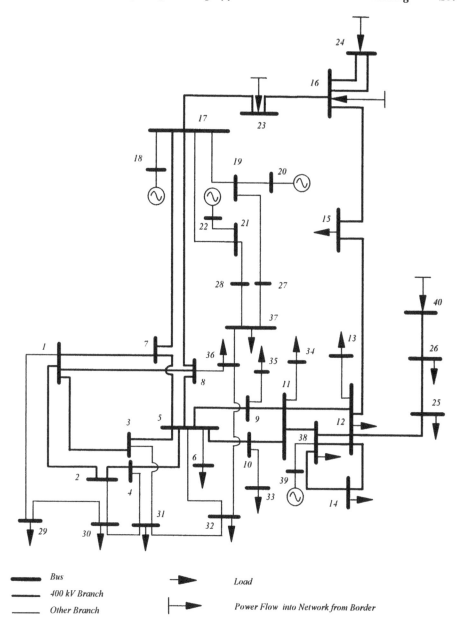

Figure 4.2 A practical UK power system network

5

Optimal Reactive Power Dispatch Using Evolutionary Programming

This chapter presents an application of evolutionary programming (EP) to optimal reactive power dispatch (ORPD). EP selects the best regulation of generator bus voltages and transformer taps to minimise the real power loss and keep the voltages and reactive power generations in their secure limits. The proposed approach is described in detail for the modified IEEE 30-bus system.

5.1   INTRODUCTION

The purpose of optimal reactive power dispatch (ORPD) is to minimise the network real power loss and improve voltage profiles by regulating the generator bus voltage, switching on or off static VAR compensators and changing transformer tap-settings. Reactive power dispatch is a complex problem in a large-scale power system. Many methods based on linear and non-linear programming have been proposed to solve this problem [1-5]. These approaches are based on successive linearisations and the first and second differentiations of an objective function and its constraint equations. Such treatments quite often lead to a local minimum point and sometimes result in divergence. Some new methods based on artificial intelligence have recently been used in ORPD and reactive power planning to solve local minimum problems and uncertainties [6-8].

This chapter presents an application of evolutionary programming (EP) to solve an ORPD problem. EP offers a new tool for the optimisation of complex power system problems.

5.2   PROBLEM FORMULATION

List of symbols

N_E — number of branches

N_i — number of buses adjacent to bus i, including bus i

N_{PQ} — number of PQ buses

N_g — number of generator buses

N_T — number of tap-setting transformer branches

N_B — total number of buses

N_{B-1} — total number of buses, excluding slack bus

P_{loss} — network real power loss

g_k — conductance of branch k

V_i — voltage magnitude at bus i

θ_{ij} — voltage magnitude difference between bus i and bus j

P_i — active power injected into the network at bus i

Q_i — reactive power injected into the network at bus i

G_{ij}, B_{ij} — mutual conductance and susceptance between bus i and bus j

G_{ii}, B_{ii} — self conductance and susceptance of bus i

Q_{gi} — reactive power generation at bus i

T_k — tap-setting of transformer branch k

$N_{V\lim}$ — number of buses with voltages outside the limits

$N_{Qg\lim}$ — number of buses with reactive power generations outside the limits

The purpose of ORPD is to minimise the network real power loss and improve voltage profiles by regulating the generator bus voltage, switching on/off static var compensators and changing transformer tap-settings. The objective function of ORPD can therefore be expressed as follows:

$$\min P_{loss} = \sum_{\substack{k \in N_E \\ k=(i,j)}} g_k (V_i^2 + V_j^2 - 2V_i V_j \cos\theta_{ij})$$

$$\text{s.t. } 0 = P_i - V_i \sum_{j \in N_i} V_j (G_{ij}\cos\theta_{ij} + B_{ij}\sin\theta_{ij}) \quad i \in N_{B-1}$$

$$0 = Q_i - V_i \sum_{j \in N_i} V_j (G_{ij}\sin\theta_{ij} - B_{ij}\cos\theta_{ij}) \quad i \in N_{PQ} \qquad (1)$$

$$Q_{gi}^{min} \leq Q_{gi} \leq Q_{gi}^{max} \qquad i \in N_g$$

$$T_k^{\ min} \leq T_k \leq T_k^{\ max} \qquad k \in N_T$$

$$V_i^{\ min} \leq V_i \leq V_i^{\ max} \qquad i \in N_B$$

where power flow equations are used as equality constraints; reactive power generation restrictions, transformer tap-setting restrictions and bus voltage restrictions are used as inequality constraints. Transformer tap-setting T and generator bus voltages V_g are control variables so they are self-restricted. Load bus voltages V_{load} and reactive power generations Q_g are state variables, which are restricted by adding them as the quadratic penalty terms to the objective function to form a penalty function. Equation (1) is therefore changed to the following generalised objective function:

$$\min \ f_q = P_{loss} + \sum_{i \in N_{V\lim}} \lambda_{V_i}(V_i - V_i^{\lim})^2$$
$$+ \sum_{i \in N_{Qg\lim}} \lambda_{Q_{gi}}(Q_{gi} - Q_{gi}^{\lim})^2$$
$$\text{s.t. } 0 = P_i - V_i \sum_{j \in N_i} V_j(G_{ij}\cos\theta_{ij} + B_{ij}\sin\theta_{ij}) \quad i \in N_{B-1}$$
$$0 = Q_i - V_i \sum_{j \in N_i} V_j(G_{ij}\sin\theta_{ij} - B_{ij}\cos\theta_{ij}) \quad i \in N_{PQ}$$

(2)

where λ_{V_i} and $\lambda_{Q_{gi}}$ are the penalty factors that can be increased in the optimisation procedure; $V_i^{\lim}$ and $Q_{gi}^{\lim}$ are defined in the following equations:

$$V_i^{\lim} = \begin{cases} V_i^{min} & \text{if } V_i < V_i^{min} \\ V_i^{max} & \text{if } V_i > V_i^{max} \end{cases}$$
$$Q_{gi}^{\lim} = \begin{cases} Q_{gi}^{min} & \text{if } Q_{gi} < Q_{gi}^{min} \\ Q_{gi}^{max} & \text{if } Q_{gi} > Q_{gi}^{max} \end{cases}$$

(3)

It can be seen that the generalised objective function f_q is a non-linear and non-continuous function. Gradient-based conventional methods are not good enough to solve this problem.

5.3   EVOLUTIONARY PROGRAMMING (EP)

EP is different from conventional optimisation methods. It does not need to differentiate cost function and constraints. It uses probability transition rules to select generations. Each individual competes with some other individuals in a combined population of the old generation and the mutated old generation. The competition results are valued using a probabilistic rule. Winners are selected to constitute the next generation and the number of winners is the same as that of individuals in the old generation.

The general process of EP has previously been described and only the initialisation will be different here. The initialisation for ORPD is set out briefly as follows:

5.3.1 Initialisation

The initial control variable population is selected by randomly selecting $p_i=[V_{PV}{}^i, T^i]$, $i=1, 2, ..., m$, where m is the population size, from the sets of uniform distribution ranging over $[V^{min}, V^{max}]$ and $[T^{min}, T^{max}]$. The fitness score f_i of each p_i is obtained by running PQ decoupled power flow.

5.4   NUMERICAL RESULTS

In this section, the IEEE 30-bus system is used to show the effectiveness of the algorithm. The system is shown in Figure 4.1 and the network parameters of the system are given in [9]. The network consists of six generator-buses, 21 load-buses and 43 branches, of which four branches, (6,9), (6,10), (4,12) and (28,27), are under-load-tap-setting transformer branches. Two cases are studied. Case 1 is of light loads whose loads and active power generations of PV buses are the same as in [9]. Case 2 is of heavy loads whose loads and active power generations of PV buses are twice those of Case 1. The reactive power generation limits are listed in Table 5.1. Voltage and tap-setting limits are listed in Table 5.2. All power and voltage quantities are per-unit values.

5.4.1 Initial Condition

All initial generator voltages and transformer taps are set to 1.0. The loads are given as:

$$\textbf{Case 1:}\quad P_{load} = 2.834 \qquad Q_{load} = 1.262$$

$$\textbf{Case 2:}\quad P_{load} = 5.668 \qquad Q_{load} = 2.524$$

The initial generations and power losses are obtained as in Table 5.3. The variables outside their limits are listed in Tables 5.4 and 5.5. In Case 2, because of heavy loads, almost all bus voltages and reactive power generations are outside their limits.

Table 5.1 Reactive power generation limits

Bus	1	2	5	8	11	13
$Q_g{}^{max}$						
Case 1	0.596	0.48	0.6	0.53	0.15	0.155
Case 2	1.192	0.96	1.2	1.06	0.30	0.31

Table 5.1 continued

Q_g^{min}

-0.298	-0.24	-0.3	-0.265	-0.075	-0.078

Table 5.2 Voltage and tap-setting limits

V_g^{max}	V_g^{min}	V_{load}^{max}	V_{load}^{min}	T_k^{max}	T_k^{min}
1.1	0.9	1.05	0.95	1.05	0.95

Table 5.3 Generations and power losses

	P_g	Q_g	P_{loss}	Q_{loss}
Case 1	2.89388	0.98020	0.05988	-0.28180
Case 2	5.94588	3.26368	0.27788	0.73968

Table 5.4 Voltages outside limits

Case 1						
Bus	26	29	30			
V_i	0.93	0.94	0.93			
Case 2						
Bus	9	10	12	14	15	16	17
V_i	0.94	0.91	0.94	0.90	0.89	0.91	0.90
Bus	18	19	20	21	22	23	24
V_i	0.87	0.86	0.873	0.87	0.88	0.87	0.85
Bus	25	26	27	29	30		
V_i ·	0.86	0.81	0.88	0.83	0.80		

Table 5.5 Reactive power generations outside limits

Bus	1	8	13
Case 1		0.569	
Case 2	-0.402	1.497	0.439

5.4.2 Optimal Results

The optimal generator-bus voltages, transformer tap-settings, generations and power losses are obtained as in Tables 5.6 to 5.8. All the state variables are regulated back into their limits.

Table 5.6 Generator-bus voltages

Bus	1	2	5	8	11	13
Case 1	1.07	1.06	1.04	1.04	1.08	1.06
Case 2	1.10	1.09	1.05	1.05	1.10	1.09

Table 5.7 Transformer tap-settings

Branch	(6,9)	(6,10)	(4,12)	(28,27)
Case 1	0.98	1.03	1.02	1.04
Case 2	1.00	1.10	1.04	1.10

Table 5.8 Generations and power losses

	P_g	Q_g	P_{loss}	Q_{loss}
Case 1	2.88357	0.87389	0.04957	-0.38811
Case 2	5.87926	2.86863	0.21126	0.34463

The power savings are given as follows:

Case 1:

$$P_{save}\% = \frac{P_{loss}^{init} - P_{loss}^{opt}}{P_{loss}^{init}} \times 100$$

$$= \frac{0.05988 - 0.04957}{0.05988} \times 100 = 17.22\%$$

Case 2:

$$P_{save}\% = \frac{P_{loss}^{init} - P_{loss}^{opt}}{P_{loss}^{init}} \times 100$$

$$= \frac{0.27788 - 0.21126}{0.27788} \times 100 = 23.97\%$$

5.5   CONCLUSIONS

ORPD is an optimisation problem of a non-continuous and non-linear function with uncertainties arising from large-scale power systems. EP with the techniques developed in this chapter is a suitable algorithm to solve such a problem. Simulation studies were carried out in a modified IEEE 30-bus system. EP performs very well in such a system and gives satisfactory results. The simulation shows that the application of EP to ORPD could achieve very attractive power savings.

5.6   ACKNOWLEGMENTS

The author wishes to acknowledge the IEEE for granting permission to reproduce the material and results contained in [10].

5.7   REFERENCES

[1]   K Mamandur and R Chenoweth, 'Optimal control of reactive power flow for improvements in voltage profiles and for real power loss minimisation', *IEEE Transactions on Power Apparatus and Systems*, **100**, 1981, 3185-3194.

[2]   M O Mansour and T M Abdel-Rahman, 'Non-linear var optimisation using decomposition and coordination', *IEEE Transactions on Power Apparatus and Systems*, **103**, 1984, 246-255.

[3]   J Qiu and S M Shahidehpour, 'A new approach for minimizing power losses and improving voltage profile', *IEEE Transactions on Power Systems*, **2**, 1987, 287-295.

[4]   O Alsac, J Bright, M Prais and B Stott, 'Further developments in LP-based optimal power flow', *IEEE Transactions on Power Systems*, **5**, 1990, 697-711.

[5]   N Deeb and S M Shahidehpour, 'Linear reactive power optimization in a large power network using the decomposition approach', *IEEE Transactions on Power Systems*, **5**, 1990, 428-438.

[6]   G Boone and H D Chiang, 'Optimal capacitor placement in distribution systems by genetic algorithm', *Electrical Power and Energy Systems*, **15**, 1993, 155-162.

[7]   Y T Hsiao, C C Liu, H D Chiang and Y L Chen, 'A new approach for optimal VAR sources planning in large scale electric power systems', *IEEE Transactions on Power Systems*, **8**, 1993, 988-996.

[8]   K H Abdul-Rahman and S M Shahidehpour, 'Application of fuzzy sets to optimal reactive power planning with security constraints', *IEEE Transactions on Power Systems*, **9**, 1994, 589-597.

[9]   K Y Lee, Y M Park and J L Ortiz, 'A united approach to optimal real and reactive power dispatch', *IEEE Transactions on Power Apparatus and Systems*, **104**, 1985, 1147-1153.

[10] J T Ma and L L Lai, 'Optimal reactive power dispatch using evolutionary programming', *IEEE/KTH Stockholm Power Tech International Symposium on Electric Power Engineering*, 1995, 662-667.

6

Application of Evolutionary Programming to Transmission Network Planning

This chapter presents an application of evolutionary programming (EP) to solve long-term transmission network planning (TNP). The non-convexity that has been observed in the network expansion planning cannot be solved effectively by conventional linear or non-linear programming. EP can discover a global or near-global optimal point in a non-smooth and non-continuous area. The simulation studies in an example system show the effectiveness of the algorithm. EP has the ability to find the global optimum point in such a non-convex function.

6.1 INTRODUCTION

It is recognised that the allocation of transmission costs in a competitive environment requires careful evaluation of alternative transmission expansion plans. As a result, the need for methods that are able to synthesise optimal transmission expansions plans has become more important than ever. Unfortunately, practice has shown that conventional optimisation procedures are unable to produce optimal solutions for networks such as those studied in [1, 2]. The reason is that the long-term transmission expansion planning problem is a hard, large-scale combinatorial problem. The number of options to be analysed increases exponentially with the size of the network. There is a larger number of local optimal solutions (a highly multimodal landscape), which makes the chances of being trapped in one of them very high.

  The objective of TNP is to determine the installation plans of new facilities (lines and other network equipment) so that the resulting bulk power system may be able to meet the forecasted demand at the lowest cost, while satisfying prescribed technical, financial and reliability criteria. This process is typically broken down into the following two stages: (i) long-term transmission expansion planning with a horizon of 15 to 30 years, and (ii) mid-term transmission expansion planning with a horizon of up to 10 years; typically 5 to 10 years. In the long-term stage, the objective is to evaluate and meet the total load demands at the lowest investment cost for network expansion, so as to establish the basic guidelines for the future network structure, while leaving a number of details to the mid-term planning, e.g. those concerning transient stability limits, voltage violations, reactive power flow and short-circuit capacity. The long-term planning has been addressed by heuristic models [1, 2], which are based on sensitivity analysis, linear programming [3], and Bender's decomposition technique and hierarchical approach [4-6]. Although these methods are successful in transmission expansion planning, some problems still exist:

1. Due to the non-convexity existing in the network expansion planning problem, the success of the search in the sequence of hyperplanes still largely depends on the starting points. Therefore, the optimisation process sometimes stops at non-optimal solutions.
2. The non-linearity of the problem increases the iterations of the optimisation algorithm and sometimes causes divergence.

As there are no fractional transmission lines, transmission expansion planning becomes a very complex mixed integer non-linear programming problem. Previous papers have shown that mathematical programming methods based on Bender's decomposition and binary search are effective in obtaining optimal solutions for small and medium size problems. Pereira and Pinto [2] showed that simulated annealing was able to provide better solutions at costs that had never before been obtained. This is the main reason for continuing to investigate the family of methods based on intelligent approach. This chapter presents an application of evolutionary programming (EP) to solve the TNP problem. EP is used to select the optimal new transmission lines and active power generations to minimise the investment cost, while meeting the total load demands without any load curtailment. The chapter is organised as follows: the TNP problem is formulated in Section 6.2; the EP for TNP is introduced in Section 6.3; the numerical results are given in Section 6.4; and the conclusions are given in Section 6.5.

6.2   PROBLEM FORMULATION

List of Symbols

P_{dicur} load curtailment at bus i (pu)

θ_{ij} voltage angle difference between buses i and j (rad)

B_{ij} mutual susceptance between buses i and j (pu)

B_{ii} self-susceptance of bus i (pu)

N_{EPlim} number of transmission lines through which the active power flow violate their limits

$N_{E\theta lim}$ number of transmission lines at which the voltage angle difference between the two terminal buses violates their limits

The objective of long-term transmission system expansion planning is to evaluate and meet the total load demands at the lowest investment cost for network expansion. Investment costs are functions of the decision variables (integer variables representing the addition of new transmission equipment); penalties are functions of the continuous operation variables (power flows). Therefore, the objective function is to minimise the investment cost and the load curtailment, subject to the system constraints as follows:

$$min\ f = \sum_{k(i,j) \in N_{Enew}} C_{1ij} n_{ij} + \sum_{i \in N_B} C_{Ci} P_{dicur}$$

$$s.t.\ 0 = P_{gi} + P_{dicur} - P_{di} - \sum_{j \in N_i} B_{ij} \theta_{ij} \quad i \in N_{B-1}$$

$$|P_k| \le P_k^{max} \qquad k \in N_E$$

$$P_{gi}^{min} \le P_{gi} \le P_{gi}^{max} \qquad i \in N_g \tag{1}$$

$$|\theta_{ij}| \le 90° \qquad k(i,j) \in N_E$$

$$0 \le n_{ij} \le n_{ij}^{max} \qquad k(i,j) \in N_{Enew}$$

where the first term in the objective function is the investment cost, which is the cost of the newly added transmission lines; the second term is the cost of load curtailments; d.c. power flow is used as an equality constraint; the active power flow limits in the transmission lines, the generation restrictions, the voltage angle difference limits of the transmission lines and the maximum new line numbers are used as inequality constraints. Since the load curtailment P_{dicur} is a state variable that cannot be fixed before the computation of the power flow, the inclusion of P_{dicur} in the power flow equation makes the equation inconvenient to compute. Therefore, in this chapter, P_{dicur} will not be used explicitly but will be expressed implicitly in the transmission line limits; that is, when the transmission line limits are violated, the amount of the violation will be the load curtailment at the receiver bus. Equation (1) is therefore changed to the following equation:

$$min f = \sum_{k(i,j) \in N_{Enew}} C_{1ij} n_{ij}$$

$$s.t.\ 0 = P_{gi} - P_{dicur} - \sum_{j \in N_i} B_{ij} \theta_{ij} \quad i \in N_{B-1}$$

$$|P_k| \le P_k^{max} \qquad k \in N_E$$

$$P_{gi}^{min} \leq P_{gi} \leq P_{gi}^{max} \qquad i \in N_g$$

$$|\theta_{ij}| \leq 90° \qquad k(i,j) \in N_E \qquad (2)$$

$$0 \leq n_{ij} \leq n_{ij}^{max} \qquad k(i,j) \in N_{Enew}$$

In the inequality constraints, the active power generations, P_{gi}, except the generation at the slack bus, P_{gs}, and the numbers of the newly added transmission lines are control variables, so they are self-restricted. The active power generation at the slack bus, P_{gs}, the active power flows in the transmission lines and the voltage angle differences are state variables, which cannot be restricted by themselves. These state variables are added as quadratic penalty terms to the objective function to form a penalty function. Since EP could deal with a non-linear function, it does not matter that the penalty function changes the original linear function to a non-linear one. Equation (2) is therefore changed to the following generalised objective function:

$$\min F = f + \sum_{k \in N_{EPlim}} \lambda_{P_k} (|P_k| - P_k^{max})^2$$

$$+ \sum_{k(i,j) \in N_{E\theta lim}} \lambda_{\theta_{ij}} (|\theta_{ij}| - 90°)^2$$

$$+ \lambda_{P_{gs}} (P_{gs} - P_{gs}^{lim})^2 \qquad (3)$$

$$\text{s.t. } 0 = P_{gi} - P_{dicur} - \sum_{j \in N_i} B_{ij}\theta_{ij} \quad i \in N_{B-1}$$

where λ_{P_k}, $\lambda_{\theta_{ij}}$ and $\lambda_{P_{gs}}$ are the penalty factors that can be increased in the optimisation procedure; P_{gs}^{lim} is defined in the following equation:

$$P_{gs}^{lim} = \begin{cases} P_{gs}^{min} & \text{if } P_{gs} < P_{gs}^{min} \\ P_{gs}^{max} & \text{if } P_{gs} > P_{gs}^{max} \end{cases} \qquad (4)$$

The penalty parameter is used to penalise the loss of load in the objective function. As a general rule, the value is such that loss of load is less attractive than investment in new transmission equipment. Towards the end of the optimisation process, the value should be high enough to discourage any type of loss of load. During the initial stages (initial populations), however, allowing some degree of loss of load is very effective in the sense that it makes it easier for the process to move through the search space. Therefore, in order to keep a larger degree of variation during the initial phases of the simulated evolution process, it is important to allow some configurations with a certain degree of loss of load.

6.3 EVOLUTIONARY PROGRAMMING

EP is different from conventional optimisation methods. It does not need to differentiate cost function and constraints. It uses probability transition rules to select generations. Each individual competes with some other individuals in a combined population of the old generation and the mutated old generation. The competition results are valued using a probabilistic rule. The same number of winners as the individuals in the old generation constitute the next generation. The procedure of EP for TNP is briefly listed as follows:

6.3.1 Initialisation

The initial control variable population is selected by randomly selecting $p_i=[n_k^i, P_g^i]$, $i=1, 2, ..., m$, where m is the population size, from the sets of uniform distribution ranging over $[n_k^{min}, n_k^{max}]$ and $[P_g^{min}, P_g^{max}]$, where $n_k=\{n_{ij} \mid k(i,j) \in N_{Enew}\}$ and $P_g=\{P_{gi} \mid i \in N_g$ and $i \neq s\}$. The fitness value f_i of each p_i is obtained by running a d.c. power flow.

The rest of the EP procedure follows the pattern of Statistics, Outer Loop and Inner Loop Start, Mutation and Competition.

For the Inner Loop Convergence Criterion, the convergence is achieved when either the maximum fitness value converges to the minimum fitness value or the generations reach the maximum generation number. If the condition is met, the process will go to the next step; otherwise, the process will go back to the Inner Loop Start.

For the Outer Loop Convergence Criterion, the convergence is achieved when either all state variables, the active power generation at the slack bus, the active power flows in the transmission lines and the voltage angle differences between the two terminal buses of the transmission lines, are within their limits or the outer loops reach the maximum number. If the condition is met, the program will stop. If one or more state variables violate their limits, the penalty factors of these variables will increase, and the process will go back to the Outer Loop Start.

6.4 NUMERICAL RESULTS

The test system is a six-bus system, which has been used in [4]. The generations and loads are given in Table 6.1 and the parameters are given in Table 6.2.

Table 6.1 Generation and load

Bus	Generation Capacity (MW)	Load (MW)
1	150	80
2		240

Table 6.1 continued

3	360	40
4		160
5		240
6	600	

Table 6.2 Line parameters

Line	Reactance (Ohm)	Transmission Limit (MW)	Investment Cost (M$)
1-2	0.40	100	40
1-3	0.38	100	38
1-4	0.60	80	60
1-5	0.20	100	20
1-6	0.68	70	68
2-3	0.20	100	20
2-4	0.40	100	40
2-5	0.31	100	31
2-6	0.30	100	30
3-4	0.59	82	59
3-5	0.20	100	20
3-6	0.48	100	48
4-5	0.63	75	63
4-6	0.30	100	30
5-6	0.61	78	61

The initial network is given in Figure 6.1. Bus 6 is a new power plant to be connected to the network, so initially there is no existing transmission line between Bus 6 to any bus in the network. Therefore, there is a power shortage of 250 MW. In the worst case, Bus 3 can at most supply 240 MW because of the transmission line limits, and the power shortage will be 370 MW instead of 250 MW.

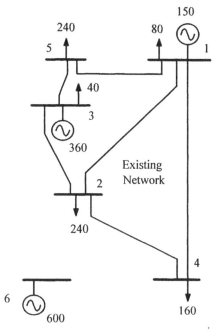

Newly Built Power Plant

Figure 6.1 The initial network

The planning allows no more than three transmission lines added between each pair of existing buses and no more than four transmission lines between the new bus, Bus 6, and each existing bus. The optimisation is carried out after three outloops. The power flow results after the outloops are given in Figures 6.2-6.4. The numbers in the brackets are transmission line limits. The transmission line power flow violation means that the load curtailment will happen at the receiver bus. The results after each outloop are given as follows:

- **Outloop 1:** One new line is added between Buses 6 and 4 as in Figure 6.2. The investment cost is 30 million dollars. The generations are 150 MW, 360 MW and 250 MW at Buses 1, 3 and 6, respectively. Three transmission lines are overloaded, which are Lines 2-3, 3-5 and 6-4. It means that there will be load curtailments of 50.323 MW at Bus 2, 150 MW at Bus 4 and 69.677 MW at Bus 5.
- **Outloop 2:** Three new transmission lines are added to the network, of which one is between Bus 3 and Bus 5, and two are between Buses 6 and 4 as in Figure 6.3. The investment cost is 80 million dollars. The generations are the same as those after Outloop 1. Three transmission lines are overloaded, which are Lines 2-3 and 6-2 (double lines). Therefore, there will be load curtailments of 28.727 MW at Bus 2 and 50 MW at Bus 4.
- **Outloop 3:** Four new transmission lines are added to the network, of which one is between Bus 3 and Bus 5, and three are between Buses 6 and 4 as in Figure 6.4. The investment cost is 110 million dollars. The generations are 150

MW, 311.344 MW and 298.656 MW at Buses 1, 3 and 6, respectively. All load curtailments are eliminated.

The results are clearly shown in Table 6.3.

Table 6.3 Optimisation results

Outloop	Load Curtailment (MW)			Investment (M$)
	Bus 2	Bus 4	Bus 5	
1	50.323	150	69.677	30
2	28.727	50	0	80
3	0	0	0	110

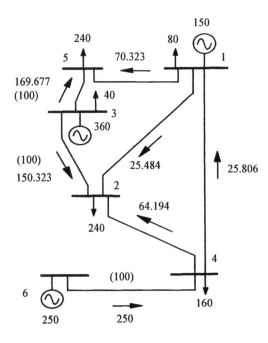

Figure 6.2 Network after Outloop 1

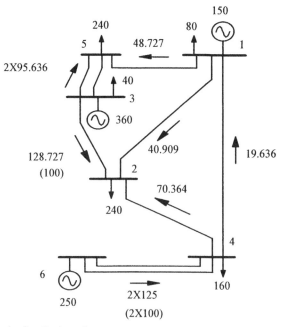

Figure 6.3 Network after Outloop 2

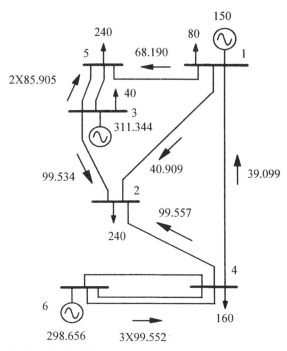

Figure 6.4 Network after Outloop 3

6.5   CONCLUSIONS

The potential for the application of EP to TNP has been shown in this chapter. TNP is a non-convexity function that is difficult to solve by conventional methods. EP is the optimisation algorithm that has the advantage in solving such a non-convexity function. The results show that EP is good at network expansion planning. EP is capable of solving the TNP problem.

6.6   ACKNOWLEGMENTS

The author wishes to acknowledge the IEE of Japan for granting permission to reproduce the material and results contained in [7].

6.7   REFERENCES

[1]   A Monticelli, A Santos Jr, M V F Pereira, S H Cunha, B J Parker and J G G Praca, 'Interactive transmission network planning using a least-effort criterion', *IEEE Transactions on Power Apparatus and Systems*, **101**, 1982, 3919-3925.

[2]   M V F Pereira and L M V G Pinto, 'Application of sensitivity analysis of load supplying capability to interactive transmission expansion planning', *IEEE Transactions on Power Apparatus and Systems*, **104**, 1985, 381-389.

[3]   R Villasana, L L Garver and S J Salon, 'Transmission network planning using linear programming', *IEEE Transactions on Power Apparatus and Systems*, **104**, 1985, 349-356.

[4]   R Romero and A Monticelli, 'A hierarchical decomposition approach for transmission network expansion planning', *IEEE Transactions on Power Systems*, **9**, 1994, 373-380.

[5]   G Latorre-Bayona and I J Perez-Arriaga, 'CHOPIN, a heuristic model for long term transmission expansion planning', *IEEE Transactions on Power Systems*, **9**, 1994,1886-1894.

[6]   R Romero and A Monticelli, 'A zero-one implicit enumeration method for optimizing investments in transmission expansion', *IEEE Transactions on Power Systems*, **9**, 1994, 1385-1391.

[7]   L L Lai, J T Ma, K P Wong, M Zhao and H Sasaki, 'Application of evolutionary programming to transmission system planning', *IEE of Japan Power & Energy '96*, 1996, 147-152.

7

Application of Evolutionary Programming to Generator Parameter Estimation

This chapter presents an artificial intelligence approach of using evolutionary programming (EP) to estimate the transient and subtransient parameters of a generator under normal operation. The estimation using EP is compared with that using corrected extended Kalman filter. The comparisons with both simulation and micromachine test results show that EP is robust in searching for the real values of parameters even when the data are highly contaminated by noise, while with extended Kalman filter, the estimation tends to diverge with such data.

7.1 INTRODUCTION

System parameter estimation techniques have been widely studied for system modelling and control. The observed stimulus-response data are usually used to estimate the parameters of the system. The objective function to be minimised is typically a function of the squared predictive error. However, this quadratic mapping of the predictive error is generally a complex, non-linear, possibly non-convex function of the parameter errors. Conventional estimation methods use the derivative of the objective function with respect to the parameters as their search directions. Such gradient-based search estimation algorithms may become trapped at local minimum points that produce an inadequate model performance. In addition, such algorithms tend to diverge when used in a real system with measured data full of noise. In general, typical identification techniques are not robust.

Estimation of dynamic parameters of generators has been a challenging problem in power systems. The accurate determination of generator parameters as operating conditions change is important for power system analysis, control system design and machine fault diagnosis. The operational behaviour of a generator is decided by its dynamic parameters and so is the behaviour of the power system. Therefore, the accuracy of power system stability analysis depends mainly on the accuracy of generator parameters. Various methods of parameter estimation have been adopted in the estimation of generator parameters [1-6]. All of the methods that have been used in power system identification are gradient-based algorithms, which are not robust, as stated above.

The estimation of an engineering system is a kind of optimisation problem that usually finds a statistical minimum of an estimation function. A wide range of efficient methods is available for the solution of problems in which the functions to be estimated are smooth and have no local minima. These methods do not work well when the function to be estimated does not have these nice properties owing to measurement noise and model uncertainties. This type of less well-behaved function exists in most engineering problems so the devising, testing and refining of new techniques for finding optimal estimation solutions are important research areas. A family of naturally inspired algorithms for optimisation that has not yet been widely used in engineering is that of the so-called evolutionary algorithms (EAs), which include genetic algorithms (GAs) and simulated evolution, evolutionary programming (EP). This chapter proposes an application of EP to estimating the parameters of a generator under normal operation. EP offers a new tool for system identification.

The generator model used in this chapter is based on Park's direct-axis and quadrature-axis representation. The parameters to be estimated have definite physical meanings, which are suitable for stability analysis and control of power systems. A small inverse pseudorandom binary sequence (PRBS) is input to the automatic voltage regulator (AVR) to disturb the generator slightly. The responses of angular speed, terminal voltage and active power of the generator are measured and the squared predictive error of responses is used as the objective function to be minimised. The estimation is updated on-line as EP proceeds so that it will be suitable for the parameter estimation of the on-line operating generators.

7.2 THE GENERATOR MODEL IN A POWER SYSTEM

List of symbols

Δ increment of the variable

δ torque angle (rad)

ω angular speed (pu)

ω_b base angular frequency (314.15926 rad / s)

X_d direct - axis synchronous reactance (pu)

X'_d direct - axis transient reactance (pu)

X''_d direct - axis subtransient reactance (pu)

X_q quadrature - axis synchronous reactance (pu)

X''_q quadrature - axis subtransient reactance (pu)

T''_{do} direct - axis, open circuit subtransient time constant (s)

T''_{qo} quadrature - axis, open circuit subtransient time constant (s)

M inertia time constant (s)

D damping coefficient (pu)

S_d saturation factor

E'_q quadrature - axis voltage behind transient reactance (pu)

E''_q quadrature - axis voltage behind subtransient reactance (pu)

E''_d direct - axis voltage behind subtransient reactance (pu)

E_{fd} field voltage (pu)

V bus voltage (pu)

u inverse PRBS input signal to automatic voltage regulator (AVR) (pu)

K_E gain of the AVR system

T_E time constant of AVR system (s)

The subtransient model of a generator with the simplified AVR is given in the following equation:

$$\Delta \dot{\delta}_i = \omega_b \Delta \omega_i$$

$$\Delta \dot{\omega}_i = -\frac{D_i}{M_i} \Delta \omega_i - \frac{I_{qi0}}{M_i} \Delta E''_{qi} + \frac{I_{di0}}{M_i} \Delta E''_{di}$$

$$-\frac{V_{di0} - X''_{di} I_{qi0}}{M_i} \Delta I_{di} - \frac{V_{qi0} + X''_{qi} I_{di0}}{M_i} \Delta I_{qi}$$

$$\Delta \dot{E}'_{qi} = -\frac{1}{T'_{doi}} \Delta E'_{qi} + \frac{1}{T'_{doi}} \Delta E_{fdi} - \frac{X_{di} - X'_{di}}{T'_{doi}} \Delta I_{di}$$

$$\Delta \dot{E}''_{qi} = -\frac{1}{T''_{doi}} \Delta E''_{qi} + \left(\frac{1}{T''_{doi}} - \frac{1+S_{di}}{T'_{doi}} \right) \Delta E'_{qi}$$

$$+ \frac{1}{T'_{doi}} \Delta E_{fdi} - \left(\frac{X'_{di} - X''_{di}}{T''_{doi}} + \frac{X_{di} - X'_{di}}{T'_{doi}} \right) \Delta I_{di}$$

$$\Delta \dot{E}''_{di} = -\frac{1}{T''_{qoi}} \Delta E''_{di} - \frac{X_{qi} - X''_{qi}}{T''_{qoi}} \Delta I_{qi}$$

$$\Delta \dot{E}_{fdi} = -\frac{1}{T_E} \Delta E_{fdi} + \frac{K_E}{T_E} (u - \Delta V_i)$$

(1)

Figure 7.1 shows the schematic diagram of a micromachine system. With V_i, I_i and V_j being measured, the following dynamic equation and its measurement equation are used in the identification program, instead of the unmeasurable I_{di} and I_{qi}:

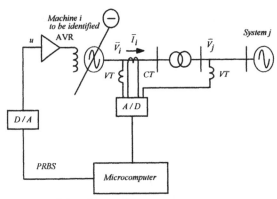

Figure 7.1 Schematic diagram of the micromachine system

$$\Delta \dot{\delta}_i = \omega_b \Delta \omega_i$$

$$\Delta \dot{\omega}_i = -\frac{K_{1i}}{M_i} \Delta \delta_i - \frac{D_i}{M_i} \Delta \omega_i - \frac{K_{2i}}{M_i} \Delta E''_{qi} - \frac{K_{3i}}{M_i} \Delta E''_{di} - \frac{C_{1i}}{M_i} \Delta V_j$$

$$\Delta \dot{E}'_{qi} = -\frac{K_{4i}}{T'_{doi}} \Delta \delta_i - \frac{1+S_{di}}{T'_{doi}} \Delta E'_{qi} - \frac{K_{5i}}{T'_{doi}} \Delta E''_{qi}$$

$$- \frac{K_{6i}}{T'_{doi}} \Delta E''_{di} + \frac{1}{T'_{doi}} \Delta E_{fdi} - \frac{C_{2i}}{T'_{doi}} \Delta V_j$$

$$\Delta \dot{E}''_{qi} = -K_{7i} \Delta \delta_i + \left(\frac{1}{T''_{doi}} - \frac{1 + S_{di}}{T'_{doi}} \right) \Delta E'_{qi} - K_{8i} \Delta E''_{qi}$$

$$- K_{9i} \Delta E''_{di} + \frac{1}{T'_{doi}} \Delta E_{fdi} - C_{3i} \Delta V_j$$

$$\Delta \dot{E}''_{di} = -\frac{K_{10i}}{T''_{qoi}} \Delta \delta_i - \frac{K_{11i}}{T''_{qoi}} \Delta E''_{qi} - \frac{K_{12i}}{T''_{qoi}} \Delta E''_{di} - \frac{C_{4i}}{T''_{qoi}} \Delta V_j$$

$$\Delta \dot{E}_{fdi} = -\frac{K_E K_{13i}}{T_E} \Delta \delta_i - \frac{K_E K_{14i}}{T_E} \Delta E''_{qi} - \frac{K_E K_{15i}}{T_E} \Delta E''_{di} \qquad (2)$$

$$- \frac{1}{T_E} \Delta E_{fdi} - \frac{K_E C_{5i}}{T_E} \Delta V_j + \frac{K_E}{T_E} u$$

$$\Delta \omega_i = \Delta \omega_i$$

$$\Delta V_i = K_{13i} \Delta \delta_i + K_{14i} \Delta E''_{qi} + K_{15i} \Delta E''_{di} + C_{5i} \Delta V_j \qquad (3)$$

$$\Delta P_i = K_{1i} \Delta \delta_i + K_{2i} \Delta E''_{qi} + K_{3i} \Delta E''_{di} + C_{1i} \Delta V_j$$

where i is the generator bus number; j is the bus after the transformer; and P_i is derived from V_i and I_i. All the coefficients are listed in the Appendix.

7.3   THE EP PARAMETER ESTIMATION ALGORITHM

To describe the estimation algorithm, Equations (2) and (3) are rewritten as follows:

$$\dot{x}(t) = A(p)x(t) + B(p)u(t) + B_e(p)y_e(t) + w(t)$$

$$y(t_k) = C(p)x(t_k) + C_e y_e(t_k) + v(t_k) \qquad (4)$$

where

$$p = \begin{bmatrix} X_{di} & X'_{di} & X''_{di} & X_{qi} & X''_{qi} & T'_{doi} & T''_{doi} & T''_{qoi} & M_i & D_i & S_{di} \end{bmatrix}^T$$

$$x(t) = \begin{bmatrix} \Delta \delta_i & \Delta \omega_i & \Delta E'_{qi} & \Delta E''_{qi} & \Delta E''_{di} & \Delta E_{fdi} \end{bmatrix}^T$$

$$y(t_k) = \begin{bmatrix} \Delta \omega_i & \Delta V_i & \Delta P_i \end{bmatrix}^T \qquad (5)$$

$$y_e(t_k) = \Delta V_j$$

p is the parameter vector to be identified; $y(t_k)$ is the measurement vector at time t_k; $y_e(t_k)$ is an extra measurement that is used to decouple the influence of the power system on the machine. $u(t)$ is a reverse PRBS input into the AVR. $w(t)$ and $v(t_k)$ are the system and measurement noises, respectively.

The procedure of EP parameter estimation is given as follows.

7.3.1 Initialisation

An initial population of parameter vectors, $\hat{p}^0 = \{\hat{p}_i^0,\ i = 1, 2, ..., m\}$, where m is the population size, are randomly selected from the sets of uniform distribution ranging over $\left[\hat{p}^{\min}, \hat{p}^{\max}\right]$, where $\hat{p}^{\min}$ and $\hat{p}^{\max}$ are approximately estimated with a priori knowledge. Let $l = 0$, where l is the generation number.

7.3.2 Integration

Each individual, $\hat{p}_i^l$ is used in a period of the integral process of the following equation:

$$\dot{x}(t) = A(\hat{p}_i^l)\hat{x}(t) + B(\hat{p}_i^l)u(t) + B_e(\hat{p}_i^l)y_e(t) \tag{6}$$

At each integral step t_k, the error equation is obtained as

$$e(\hat{p}_i^l, t_k) = y(t_k) - C(\hat{p}_i^l)\hat{x}(t_k) - C_e(\hat{p}_i^l)y_e(t_k) \tag{7}$$

The fitness is the weighted variance of the error e as follows:

$$f_i(\hat{p}_i^l) = E[e(\hat{p}_i^l, t_k)^T \Lambda e(\hat{p}_i^l, t_k)] \tag{8}$$

where Λ is a diagonal weight matrix, which can be a unit matrix if there is no special measurement error.

7.3.3 Mutation

Each parameter, $\hat{p}_{i,j}^l$, $i=1, ..., m$, $j=1, ..., n$, where n is the number of parameters, will add a Gaussian random variable to produce a mutated parameter, $\hat{p}_{i+m,j}^l$, in accordance with the following equation:

$$\hat{p}_{i+m,j}^l = \hat{p}_{i,j}^l + N\left(0, \beta_j \frac{f_i}{f_{\max}}\right), \quad i = 1, ..., m \quad j = 1, ..., n \tag{9}$$

where, $\hat{p}_{i,j}^l$ denotes the jth element of the ith individual; $N(\mu, \sigma^2)$ represents a Gaussian random variable with mean μ and variance σ^2; $f_{\max}$ is the maximum fitness of the old generation; β_j is a coefficient of proportionality used to scale $f_i/f_{\max}$, which is given in the following equation:

$$\beta_j = p_{\text{mut}}(\hat{p}_j^{\max} - \hat{p}_j^{\min}) \tag{10}$$

where p_{mut} is the mutation probability and $\hat{p}_j^{max} - \hat{p}_j^{min}$ defines the parameter range.

The following equation is used to guarantee the feasibility:

$$\hat{p}_{i+m,j}^l = \begin{cases} \hat{p}_{i,j}^l + N\left(0, \beta_j \dfrac{f_i}{f_\Sigma}\right), & \text{if } \hat{p}_j^{min} \leq \hat{p}_{i,j}^l + N\left(0, \beta_j \dfrac{f_i}{f_\Sigma}\right) \leq \hat{p}_j^{max} \\ \hat{p}_j^{min} & \text{if } \hat{p}_{i,j}^l + N\left(0, \beta_j \dfrac{f_i}{f_\Sigma}\right) < \hat{p}_j^{min} \\ \hat{p}_j^{max} & \text{otherwise} \end{cases} \qquad (11)$$

Each individual, $\hat{p}_{i+m}^l$, is used in the Integration process (see Section 7.3.2) to produce its fitness f_{i+m}. A combined population is formed with the old generation and the mutated old generation.

In general EP, mutation probability is fixed throughout the whole search process. However, in practical applications, a small fixed mutation probability can only result in premature convergence, while the search with a large fixed mutation probability does not converge. An adaptive mutation probability is given to solve the problem as follows:

$$p_{mut}(l+1) = \begin{cases} p_{mut}(l) - p_{mutstep} & \text{if } f_{min}(l) \text{ unchanged or even increased} \\ p_{mut}(l) & \text{if } f_{min}(l) \text{ decreased} \\ p_{mutfinal} & \text{if } p_{mut}(l) - p_{mutstep} < p_{mutfinal} \end{cases} \qquad (12)$$

$$p_{mut}(0) = p_{mutinit}$$

where $p_{mutinit}$, $p_{mutfinal}$ and $p_{mutstep}$ are fixed numbers. $p_{mutinit}$ would be around 1 and $p_{mutfinal}$ would be 0.005. $p_{mutstep}$ depends on the maximum generation number. These parameters guarantee that if the fitness (objective function) decreases, the mutation probability will be kept large for the expanded search area to make sure that even better results are not left out. On the other hand, if there is no improvement or even worsening at this step, the mutation probability will decrease to make the smooth convergence.

The rest of the EP procedure is the same as before.

7.4   RESULTS

The results are obtained with both the simulated identification and a micromachine test identification. A corrected extended Kalman filter (CEKF) generator identification program [6] is used as the comparison program.

7.4.1 Simulation Results

In the simulation, a generator is operating in a system with changing bus voltages and frequencies. The system is represented by an equivalent generator that is ten times bigger in capacity than the generator to be estimated as in Figure 7.1. Four cases are studied, which are under disturbances with the same 0.03 pu inverse PRBS input into the AVR but the responses are measured with the simulated measurement noises of different variances. The signal-to-noise ratios according to the following equation are given in Table 7.1.

$$\left[\frac{S}{N}\right]_{dB} = 10\log_{10}\left(\frac{E(s^2(t))}{E(n^2(t))}\right) \tag{13}$$

where s is the response and n is the measurement noise, $E(\cdot)$ is the mean-square value as follows:

$$E(s^2(t)) = \frac{1}{T}\sum_{t=1}^{T}s^2(t)$$
$$E(n^2(t)) = \frac{1}{T}\sum_{t=1}^{T}n^2(t) \tag{14}$$

In Case 4, the noises are quite high. However, even in such a high noise situation, EP still gives satisfactory results. The results are shown in Table 7.2. It can be seen that in no-noise or low-noise simulations, the results of CEKF are the same as, or even better than, those of EP. However, in high-noise simulations, the performance of CEKF deteriorates, while EP still gives satisfactory results. The comparison shows that the estimation using EP is robust to search the real values of parameters even with data that are highly contaminated by noise, while that with an extended Kalman filter tends to diverge with such data. The results of extensive computer simulation show that EP is a robust search program to obtain good parameter estimates from the data full of noise. Such robust characteristics are essential in parameter and state estimations in real systems.

Table 7.1 Signal-to-noise ratio

$\left[\dfrac{S}{N}\right]_{dB}$	$\Delta\omega_i$	ΔV_i	ΔP_i	ΔV_j
Case 2	38.3	63.6	71.6	40.6
Case 3	18.6	43.5	51.7	20.7
Case 4	4.27	9.49	17.90	0.63

Table 7.2 The simulated estimation results

		X_d	X_d'	X_d''	X_q	X_q''	T_{do}'	T_{do}''	T_{qo}''	M	D	S_d	Max Errors
True Values		2.0	0.244	0.185	1.91	0.212	4.18	0.75	0.743	6.5	5.012	1.0	
Case 1	EP	2.009	0.243	0.184	1.922	0.214	4.240	0.737	0.732	6.398	4.919	1.010	1.86% (D)
	CEKF	1.997	0.245	0.185	1.904	0.213	4.187	0.750	0.745	6.482	5.017	1.005	0.47% (S_d)
Case 2	EP	2.035	0.238	0.180	1.854	0.207	4.259	0.757	0.769	6.319	4.891	0.988	3.50% (T_{qo}'')
	CEKF	2.049	0.240	0.188	1.879	0.209	4.280	0.759	0.754	6.673	4.883	0.976	2.66% (M)
Case 3	EP	2.127	0.234	0.191	1.811	0.213	3.858	0.684	0.777	6.324	4.675	0.941	8.73% (T_{do}'')
	CEKF	1.894	0.257	0.223	1.450	0.251	3.329	0.928	0.737	6.100	4.468	0.889	24.08% (X_q)
Case 4	EP	1.782	0.210	0.154	2.147	0.242	3.521	0.857	0.660	6.306	5.688	0.940	16.51% (X_d'')
	CEKF	4.100	0.082	0.033	3.743	0.182	6.389	1.401	0.201	14.569	6.969	0.405	124.15% (M)

7.4.2 Micromachine Identification Results

The d.c. motor-generator micromachine with the thyristor excitation system is used as the test machine. Its ratings are: S_N=5 kVA, V_N=220 V, $\cos\varphi_N$=0.85, S_N=1500 rpm.

The fifth-order inverse PRBS of 0.05 pu magnitude is the input to the AVR. Three measurements of V_i, I_i and V_j are obtained with a microcomputer measurement system, from which ΔV_i, ΔP_i and ΔV_j are derived. $\Delta\omega_i$ are obtained from the counts per cycle in the wave of V_i, with 20000 counts per cycle as the base. The measurement noises are very high. Three different operating conditions are tested, no-load, half-load and full-load. The identification results of the full-load test are given in Table 7.3. The responses of $\Delta\omega_i$, ΔV_j and ΔP_i derived from the measurements under full-load conditions are shown in Figures 7.2-7.4, respectively, together with the responses from the system simulated with the design parameters, which were supplied by the manufacturers based on the open-circuit and sustained short-circuit tests, and with CEKF and EP estimated parameters. For presentation purposes, only the data in the range 6.2-12.4 are given in the figures. It is clearly shown that the simulated responses from the EP-estimated system are the most matched ones, followed by the responses from the CEKF-estimated system, while the ones from the system with design parameters do not match up well. The identification from the other two operating conditions shows that EP also gives better results.

Table 7.3 The micromachine test results

	X_d	X_d'	X_d''	X_q	X_q''	T_{do}'	T_{do}''	T_{qo}''	M	D	S_d
Designed Values	1.806	0.286	0.135	1.655	0.112	6.2	0.75	0.743	6	—	—
Offline Experiment	1.577	0.190	0.095	1.524	0.091	4.14	0.82	0.35	4.49	2.43	0.58
CEKF Values	1.577	0.190	0.095	1.524	0.091	4.14	0.82	0.35	4.49	2.43	0.58
EP Values	1.532	0.265	0.123	1.376	0.104	6.5	0.89	0.57	6.03	2.24	0.73

NB   All parameters given in Tables 7.2 and 7.3 are described in the list of symbols in Section 7.2.

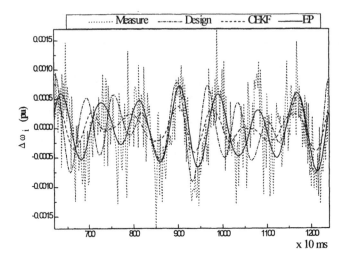

Figure 7.2 Responses of frequencies under full-load condition

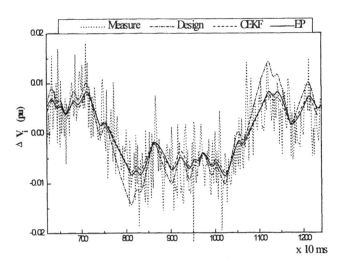

Figure 7.3 Responses of terminal voltages under full-load condition

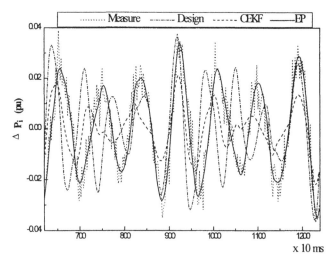

Figure 7.4 Responses of active powers under full-load condition

7.5 CONCLUSIONS

EP has been successfully applied in estimating the dynamic parameters of a synchronous generator operating in a power system. The results of extensive computer simulation and micromachine tests show that EP is a robust search program to obtain good parameter estimates from the data full of noise. Such robust characteristics are essential in real-life parameter and state estimations.

7.6 ACKNOWLEGMENTS

The author wishes to acknowledge the IEEE for granting permission to reproduce the material and results contained in [7].

7.7 REFERENCES

[1] C C Lee and Q T Tan, 'A weighted-least-squares parameter estimator for synchronous machines', *IEEE Transactions on Power Apparatus and Systems*, **PAS-96**, 1977, 97-101.

[2] M Namba, T Nishiwaki, S Yokokawa, K Ohtsuka and Y Ueki, 'Identification of parameters for power system stability analysis using Kalman filter', *IEEE Transactions on Power Apparatus and Systems*, **PAS-100**, 1981, 3304-3311.

[3] K E Bollinger, H S Khalil, L C C Li, and W E Norum, 'A method for on-line identification of power system model parameters in the presence of noise', *IEEE Transactions on Power Apparatus and Systems*, **PAS-101**, 1982, 3105-3111.

[4] N Jaleeli, M S Bourawi and J H Fish III, 'A quasilinearization based algorithm for the identification of transient and subtransient parameters of synchronous machines', *IEEE Transactions on Power Systems*, **PWRS-1**, 1986, 46-52.

[5] L X Le and W J Wilson, 'Synchronous machine parameter identification: a time domain approach', *IEEE Transactions on Energy Conversion*, **EC-3**, 1988, 241-248.

[6] J T Ma, B W Hogg, Z Y Nan and Y H Yang, 'Determination of synchronous generator parameters by system identification', *International Conference on Electrical Machines, Manchester, UK*, 1992, 54-58.

[7] L L Lai and J T Ma, 'Application of evolutionary programming to transient and subtransient parameter estimation', *IEEE Transactions on Energy Conversion*, **11**, 1996, 523-530.

7.8   APPENDIX

$$
\begin{aligned}
C_1 &= 1 + RG - XB \qquad C_2 = RB + XG \\
R_1 &= R - C_2 X_{di}^{''} \qquad R_2 = R - C_2 X_{qi}^{''} \\
X_1 &= X + C_1 X_{qi}^{''} \qquad X_2 = X + C_1 X_{di}^{''} \\
Z_i^2 &= R_1 R_2 + X_1 X_2
\end{aligned}
\tag{A1}
$$

where G and B are conductance and susceptance from Bus i to neutral, representing the local load; R and X are the resistance and reactance of the transformer:

$$
\begin{aligned}
\begin{bmatrix} F_{di} \\ F_{qi} \end{bmatrix} &= \frac{V_{i0}}{Z_i^2} \begin{bmatrix} X_1 \sin\delta_{i0} - R_2 \cos\delta_{i0} \\ R_1 \sin\delta_{i0} + X_2 \cos\delta_{i0} \end{bmatrix} \\
\begin{bmatrix} Y_{dqi} \\ Y_{qqi} \end{bmatrix} &= \frac{1}{Z_i^2} \begin{bmatrix} C_1 X_1 - C_2 R_2 \\ C_1 R_1 + C_2 X_2 \end{bmatrix} \\
\begin{bmatrix} Y_{ddi} \\ Y_{qdi} \end{bmatrix} &= \frac{1}{Z_i^2} \begin{bmatrix} -C_1 R_2 - C_2 X_1 \\ C_1 X_2 - C_2 R_1 \end{bmatrix} \\
\begin{bmatrix} G_{di} \\ G_{qi} \end{bmatrix} &= \frac{1}{Z_i^2} \begin{bmatrix} -R_2 \sin\delta_{i0} - X_1 \cos\delta_{i0} \\ X_2 \sin\delta_{i0} - R_1 \cos\delta_{i0} \end{bmatrix}
\end{aligned}
\tag{A2}
$$

$$
\begin{bmatrix} K_{1i} \\ K_{2i} \\ K_{3i} \\ C_{1i} \end{bmatrix} = \begin{bmatrix} 0 \\ I_{qi0} \\ -I_{di0} \\ 0 \end{bmatrix} + \begin{bmatrix} F_{di} & F_{qi} \\ Y_{dqi} & Y_{qqi} \\ Y_{ddi} & Y_{qdi} \\ G_{di} & G_{qi} \end{bmatrix} \begin{bmatrix} V_{di0} - X_{di}^{''} I_{qi0} \\ V_{qi0} + X_{qi}^{''} I_{di0} \end{bmatrix}
\tag{A3}
$$

$$
\begin{bmatrix} K_{4i} \\ K_{5i} \\ K_{6i} \\ C_{2i} \end{bmatrix} = \begin{bmatrix} F_{di} \\ Y_{dqi} \\ Y_{ddi} \\ G_{di} \end{bmatrix} (X_{di} - X_{di}')
\tag{A4}
$$

$$
\begin{bmatrix} K_{7i} \\ K_{8i} \\ K_{9i} \\ C_{3i} \end{bmatrix} = \begin{bmatrix} 0 \\ \dfrac{1}{T''_{doi}} \\ 0 \\ 0 \end{bmatrix} + \begin{bmatrix} F_{di} \\ Y_{dqi} \\ Y_{ddi} \\ G_{di} \end{bmatrix} \left(\dfrac{X_{di} - X'_{di}}{T'_{doi}} + \dfrac{X'_{di} - X''_{di}}{T''_{doi}} \right) \tag{A5}
$$

$$
\begin{bmatrix} K_{10i} \\ K_{11i} \\ K_{12i} \\ C_{4i} \end{bmatrix} = \begin{bmatrix} 0 \\ 0 \\ 1 \\ 0 \end{bmatrix} + \begin{bmatrix} F_{di} \\ Y_{qqi} \\ Y_{qdi} \\ G_{qi} \end{bmatrix} (X_{qi} - X''_{qi}) \tag{A6}
$$

$$
\begin{bmatrix} K_{13i} \\ K_{14i} \\ K_{15i} \\ C_{5i} \end{bmatrix} = \begin{bmatrix} 0 \\ \dfrac{V_{qi0}}{V_{i0}} \\ -\dfrac{V_{di0}}{V_{i0}} \\ 0 \end{bmatrix} + \begin{bmatrix} F_{di} & F_{qi} \\ Y_{dqi} & Y_{qqi} \\ Y_{ddi} & Y_{qdi} \\ G_{di} & G_{qi} \end{bmatrix} \begin{bmatrix} -X''_{di} \dfrac{V_{qi0}}{V_{i0}} \\ X''_{qi} \dfrac{V_{di0}}{V_{i0}} \end{bmatrix} \tag{A7}
$$

8

Evolutionary Programming for Economic Dispatch of Units with Non-Smooth Input-Output Characteristic Functions

This chapter presents an application of evolutionary programming (EP) to solve the economic dispatch (ED) of units with non-smooth input-output characteristic functions. The IEEE 30-bus system with six generating units is used as the simulation system to show the effectiveness of the algorithm. The EP finds the global or near-global optimum point in such a non-smooth function

8.1   INTRODUCTION

The process of scheduling generation to minimise the operating cost is called economic dispatch (ED). In this calculation, the generation costs are represented as curves, usually piecewise linear, and the overall calculation minimises the operating cost by finding a point where the total output of the generations equals the total power that must be delivered and where the incremental cost of power generation is equal for all generators. However, if a generator is at its upper or lower limit, that generator's incremental cost is different. Economic dispatch calculations take account of the network losses through the use of incremental loss factors. To accomplish this is to run an optimal power flow (OPF) that minimises the generation cost while taking account of the entire transmission system and all its constraints.

ED is an important daily optimisation task in the operation of a power system. Various mathematical programming methods and optimisation techniques have been applied to ED. Most of these are calculus-based optimisation algorithms that are based on successive linearisations and use the first and second differentiations of objective function and its constraint equations as the search directions [1]. They usually require the heat input-electric power output characteristics of generators to be of a monotonically increasing nature or of a piecewise linearity. However, large modern generating units with multivalve steam turbines exhibit a large variation in the input-output characteristic functions. The valve-point effects, owing to wire drawing as each steam admission valve starts to open, typically produce a ripple-like heat rate curve [2]. The conventional optimisation methods are not suitable to solve such a problem. Dynamic programming had been proposed to solve the ED problem with non-smooth generator cost curves [3]. However, the dimensions of the problem would become extremely large with the increase of the variables (curse of dimensionality). Simulated annealing (SA) [4] and genetic algorithms (GAs) [5] based on artificial intelligence have recently been proposed to solve such non-smooth ED problems. Mutalik et al. [6] have concluded, from their tested cases, that GAs perform consistently better than SA.

This chapter presents an application of evolutionary programming (EP) to solve the ED problem. The IEEE 30-bus system is used as the simulation system to show the effectiveness of the algorithm. The network consists of six generator units. The input-output curves of the generating units are represented by second-order polynomial functions, superimposed with the valve-point effects. The curves are therefore non-monotonically increasing curves with multiple local minimum points. The EP with the techniques developed in this chapter is a suitable algorithm to solve such a problem. EP finds the global optimum point in such a non-smooth function.

8.2 PROBLEM FORMULATION

List of Symbols

H heat input to the unit (Mbtu/h)
N_i number of buses adjacent to bus i, including bus i
N_g number of generator buses
N_{B-1} total number of buses, excluding the slack bus
N_{PQ} number of PQ buses
V_i voltage magnitude at bus i, (pu)
θ_{ij} voltage angle difference between bus i and bus j (rad)
P_i, Q_i active and reactive powers injected into the network at bus i, (pu)
G_{ij}, B_{ij} mutual conductance and susceptance between bus i and bus j (pu)
G_{ii}, B_{ii} self-conductance and susceptance of bus i (pu)
P_{gi} active power generation at bus i (pu)
P_g active power generation vector (pu)

P_g^i active power generation vector in generation (iteration) i of the EP process
(pu)

8.2.1 Valve Point Effect

Large steam turbine generators usually have a number of steam admission valves
that are opened in sequence to obtain ever-increasing output from the unit.
Figure 8.1 shows both input-output and incremental heat rate characteristics for a
unit with four valves. As the unit loading increases, the input to the unit
increases. However, when a valve is just opened, the throttling losses increase
rapidly so that the input heat increases rapidly and the incremental heat rate rises
suddenly. This gives rise to the non-smooth type of heat input and discontinuous
type of incremental heat rate characteristics. This type of characteristic should be
used in order to schedule steam units accurately, but it cannot be used in
traditional optimisation methods because it does not meet the convex condition.

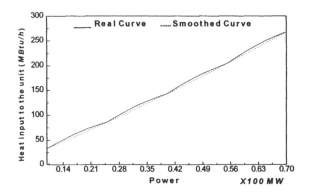

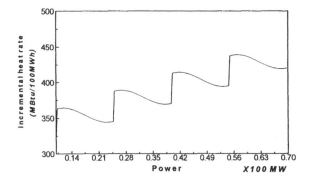

Figure 8.1 Characteristics of a steam turbine generator with four steam admission valves

8.2.2   Objective Function

To model the effects of valve points, a recurring rectified sinusoid contribution is added to the second-order polynomial functions to represent the input-output equation as follows [5]:

$$f_p = \sum_{i \in N_g} (a_i + b_i P_{gi} + c_i P_{gi}^2 + |e_i \sin(f_i (P_i - P_i^{\min}))|) \tag{1}$$

The objective function of ED is to minimise the above non-smooth function by regulating the active power outputs of the generators, subject to all operational and secure constraints. The objective function of ED is therefore expressed as follows:

$$\min f_p = \sum_{i \in N_g} (a_i + b_i P_{gi} + c_i P_{gi}^2 + |e_i \sin(f_i (P_i - P_i^{\min}))|)$$

$$\text{s.t. } 0 = P_i - V_i \sum_{j \in N_i} V_j (G_{ij} \cos\theta_{ij} + B_{ij} \sin\theta_{ij}) \quad i \in N_{B-1}$$

$$0 = Q_i - V_i \sum_{j \in N_i} V_j (G_{ij} \sin\theta_{ij} - B_{ij} \cos\theta_{ij}) \quad i \in N_{PQ} \tag{2}$$

$$P_{gi}^{\min} \le P_{gi} \le P_{gi}^{\max} \quad i \in N_g$$

where power flow equations are used as equality constraints, which represent the network power losses but are more accurate than the B matrix representation that is used in most ED programs; active power generation restrictions are used as inequality constraints. The active power generation P_{gs} at slack bus, where the subscript s represents the slack bus, is the state variable that is restricted by a quadratic penalty term added to the objective function to form a penalty function. Equation (2) is therefore changed to the following generalised objective function:

$$\min f = \sum_{\substack{i \in N_g \\ i \neq s}} (a_i + b_i P_{gi} + c_i P_{gi}^2 + |e_i \sin(f_i (P_i - P_i^{\min}))|)$$

$$+ \lambda_{P_{gs}} (P_{gs} - P_{gs}^{\lim})^2$$

$$\text{s.t. } 0 = P_i - V_i \sum_{j \in N_i} V_j (G_{ij} \cos\theta_{ij} + B_{ij} \sin\theta_{ij}) \quad i \in N_{B-1} \tag{3}$$

$$0 = Q_i - V_i \sum_{j \in N_i} V_j (G_{ij} \sin\theta_{ij} - B_{ij} \cos\theta_{ij}) \quad i \in N_{PQ}$$

$$P_{gi}^{\min} \le P_{gi} \le P_{gi}^{\max} \quad i \in N_g \quad i \neq s$$

where $\lambda_{P_{gs}}$ is the penalty factor that can be increased in the optimisation procedure; the inequality constraints are control variable constraints, so they are self-restricted; and $P_{gs}^{\lim}$ is defined in the following equation:

$$P_{gs}^{lim} = \begin{cases} P_{gs}^{min} & \text{if} \quad P_{gs} < P_{gs}^{min} \\ P_{gs}^{max} & \text{if} \quad P_{gs} > P_{gs}^{max} \end{cases} \tag{4}$$

8.3   EVOLUTIONARY PROGRAMMING (EP)

EP is different from conventional optimisation methods. It does not need to differentiate cost function and constraints. It uses probability transition rules to select generations. Each individual competes with some other individuals in a combined population of the old generation and the mutated old generation. The competition results are valued using a probabilistic rule. The same number of winners as the individuals in the old generation constitute the next generation. The procedure of EP for ED is briefly listed as follows:

8.3.1 Initialisation

The initial control variable population is selected by randomly selecting $p_i=[P_g{}^i]$, $i=1, 2, ..., m$, where m is the population size, from the sets of uniform distribution ranging over $[P_g{}^{min}, P_g{}^{max}]$. The fitness value f_i of each p_i is obtained by running a PQ decoupled power flow.

The rest of the EP procedures are the same as before.

8.4   NUMERICAL RESULTS

In this section, the modified IEEE 30-bus system is used to show the effectiveness of the algorithm. The system is shown in Figure 4.1 of Chapter 4. Six generators are used in dispatch. The parameters of the characteristics of the steam turbine generators are given in Table 8.1. The powers in the table are per-unit on a power base of

$$S_B = 100 \text{ MVA}$$

Table 8.1 Parameters of the characteristics of steam turbine generators

Generator	Unit 1	Unit 2	Unit 3
Bus	1	2	5
P_{max} (pu)	2.5	1.6	1
P_{min} (pu)	0.5	0.2	0.15

Table 8.1 continued

a (MBtu)	0	0	0
b (MBtu/puW)	200	175	100
c (Mbtu/puW^2)	37.5	175	625
e (MBtu)	15	10	10
f (rad/puW)	6.283	8.976	14.784
Generator	Unit 4	Unit 5	Unit 6
Bus	8	11	13
P_{max} (pu)	0.7	0.6	0.8
P_{min} (pu)	0.1	0.1	0.12
a (MBtu)	0	0	0
b (MBtu/puW)	325	300	300
c (MBtu/puW^2)	83.4	250	250
e (MBtu)	5	5	5
f (rad/puW)	20.944	25.133	18.48

The load periods are simply divided into the valley load duration and the peak load duration. The valley loads and the initial generations are the same as those in [17] of Chapter 4. The peak loads are twice the valley loads. The results are given in Table 8.2. The heat saved is 9.61% in the valley load duration and 14.84% in the peak load duration.

Table 8.2 Optimisation results

Population Size	40		
Valley Loads			
Initial Results			
Generator	Unit 1	Unit 2	Unit 3
Table 8.2 continued			
P_g (pu)	0.99388	0.8	0.5
Generator	Unit 4	Unit 5	Unit 6
P_g (pu)	0.2	0.2	0.2
Heat Consumed (MBtu)	902.403		
Optimal Results (486/50 CPU Time = 3.16 min)			
Generator	Unit 1	Unit 2	Unit 3
P_g (pu)	1.99007	0.50852	0.15

Table 8.2 continued

Generator	Unit 4	Unit 5	Unit 6
P_g (pu)	0.1	0.1	0.12
Heat Consumed (MBtu)	815.268		

Peak Loads

Initial Results

Generator	Unit 1	Unit 2	Unit 3
P_g (pu)	2.14588	1.6	1
Generator	Unit 4	Unit 5	Unit 6
P_g (pu)	0.4	0.4	0.4
Heat Consumed (Mbtu)	2518.200		

Optimal Results (486/50 CPU Time = 5.24 min)

Generator	Unit 1	Unit 2	Unit 3
P_g (pu)	2.50	1.24982	0.36240
Generator	Unit 4	Unit 5	Unit 6
P_g (pu)	0.7	0.6	0.58447
Heat Consumed (MBtu)	2144.451		

8.5   CONCLUSIONS

The potential for the application of EP to ED has been shown in this chapter. The results show that EP is good at non-smooth functions.

8.6   ACKNOWLEGMENTS

The author wishes to acknowledge the PSCC for granting permission to reproduce the material and results contained in [7].

8.7   REFERENCES

[1]   A J Wood and B F Wollenberg, *Power Generation, Operation & Control*, John Wiley & Sons, 1984.
[2]   IEEE Committee Report, 'Present practices in the economic operation of power systems', *IEEE Transactions on Power Apparatus and Systems*, **90**, 1971, 1768-1775.
[3]   R R Shoults, S V Venkatesh, S D Helmick, C L Ward and M J Lollar, 'A dynamic programming based method for developing dispatch curves when incremental heat rate curves are non-monotonically increasing', *IEEE Transactions on Power Systems*, **1**, 1986, 10-16.

[4] K P Wong and C C Fung, 'Simulated annealing based economic dispatch algorithm', *IEE Proceedings - Generation, Transmission and Distribution*, **140**, 1993, 509-515.

[5] D C Walters and G B Sheble, 'Genetic algorithm solution of economic dispatch with valve point loading', *IEEE Transactions on Power Systems*. **8**, 1993, 1325-1332.

[6] P P Mutalik, L R Knight, J L Blanton and R L Wainwright, 'Solving combinatorial optimisation problems using parallel simulated annealing and parallel genetic algorithms', *Proceedings of the 1992 ACM/SIGAPP Symposium on Applied Computing*, USA, 1992, 1031-1038.

[7] L L Lai, J T Ma and K P Wong, 'Evolutionary programming for economic dispatch of units with non-smooth input-output characteristic functions', *Proceedings of the twelfth Power Systems Computation Conference*, PSCC, 1996, 492-498.

9

Power Flow Control in FACTS Using Evolutionary Programming

This chapter presents the use of an evolutionary programming (EP) to solve optimal power flow (OPF) problems in flexible AC transmission systems (FACTS). The unified power flow controller (UPFC) is used as a phase shifter and/or series compensator to regulate both the angles and magnitude of branch voltages. EP, coupled with a PQ power flow, selects the best regulation to minimise the real power loss and keep the power flows within their secure limits.

9.1 INTRODUCTION

FACTS (flexible AC transmission systems) with the new technology of power electronics has given new control facilities in power systems, in both steady-state power flow control and dynamic stability control. In the steady-state operation of power systems, unwanted loop power flow and parallel power flow between utilities are problems in heavily interconnected bulk power systems. These two power flows are sometimes beyond the control of generators or they may cost too much because of generator regulations. However, with the phase shifter and/or other control facilities based on fast reacted power electronic components in networks, the unwanted power flow will be easily regulated. Several papers have been published on dealing with power flow controls [1-4]. The UPFC with voltage source converters can operate as a shunt compensator, tap-changer, series compensator and phase shifter. It is possible to make use of circuit reactances and voltage angles as control facilities for power flows in the network. However, extra control facilities complicate the operation of the system. As the control facilities influence each other, good co-ordination is required in order to make all the devices to work together, without interfering with each other. Therefore, it becomes necessary to extend the available system analysis tools, such as the optimal power flow (OPF), to represent FACTS controls. It has also been noted that the OPF problem with series compensation may be a non-convex problem [1], which could lead the conventional optimisation methods to be stuck in local minima.

This chapter presents an application of EP to solving the OPF problems in FACTS. Simulation studies are carried out in a modified IEEE 30-bus system. The phase shifters and/or series compensators based on UPFC are adopted as the power flow control facilities. The active power loss is used as the objective function and the active power flow limits in transmission lines during the contingency state are used to justify the UPFC control.

The objectives for optimal power flows are to minimise production costs, MW losses or MVAR losses and control variable shift. The control variables could be phase shift transformer taps, switched capacitors, SVC control setting, generator MW, voltage and MVAR control, load shedding and line switching.

9.2   FACTS MODEL

9.2.1 Phase Shifter

The power injection model is used to represent the phase shifter, which is derived in [2-4] but is based on the phase shifter transformer. A general model based on UPFC is given in this chapter. The tap change is independent of the phase shift so that the angle and the magnitude of the transmission line voltage can be regulated separately. The equivalent circuit of the model is given in Figure 9.1. A variable with a bar above represents a complex number.

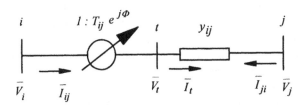

Figure 9.1 Equivalent circuit of power flow controller

The power flow equations of the branch can be derived as follows:

$$P_{ij} = V_i^2 T_{ij}^2 g_{ij} - V_i V_j T_{ij}(g_{ij}\cos(\theta_{ij} + \Phi) + b_{ij}\sin(\theta_{ij} + \Phi))$$

$$Q_{ij} = -V_i^2 T_{ij}^2 b_{ij} - V_i V_j T_{ij}(g_{ij}\sin(\theta_{ij} + \Phi) - b_{ij}\cos(\theta_{ij} + \Phi))$$

$$P_{ji} = V_j^2 g_{ij} - V_i V_j T_{ij}(g_{ij}\cos(\theta_{ij} + \Phi) - b_{ij}\sin(\theta_{ij} + \Phi))$$

$$Q_{ji} = -V_j^2 b_{ij} + V_i V_j T_{ij}(g_{ij}\sin(\theta_{ij} + \Phi) + b_{ij}\cos(\theta_{ij} + \Phi))$$

(1)

where

$$y_{ij} = g_{ij} + jb_{ij}$$
$$\overline{V}_i = V_i e^{j\theta_i} \tag{2}$$
$$\theta_{ij} = \theta_i - \theta_j$$

The effect of the phase shifter can be represented by equivalent injected powers at both buses of the branch as in Figure 9.2. and a general power injection model based on UPFC is given in (3). Considering that g_{ij} is much less than b_{ij}, the following equation could be derived:

$$P_{is} = b_{ij}V_iV_j(T_{ij}\sin(\theta_{ij} + \varPhi) - \sin\theta_{ij})$$
$$Q_{is} = b_{ij}(V_i^2(T_{ij}^2 - 1) - V_iV_j(T_{ij}(\cos(\theta_{ij} + \varPhi) - \cos\theta_{ij})))$$
$$P_{js} = -b_{ij}V_iV_j(T_{ij}\sin(\theta_{ij} + \varPhi) - \sin\theta_{ij}) \tag{3}$$
$$Q_{js} = -b_{ij}V_iV_j(T_{ij}(\cos(\theta_{ij} + \varPhi) - \cos\theta_{ij}))$$

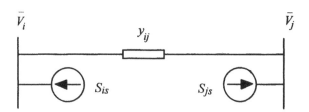

Figure 9.2 Power injection model

These equivalent injected powers, together with a regular transmission branch as in Figure 9.2, are used in the power flow equations. The derivation in [3] shows that the sensitivities of the equivalent injected powers to the magnitudes and angles of voltages at Buses i and j are much smaller than the corresponding elements in the ordinary Jacobian matrix of power flow equations. Therefore, these injected powers can be treated as loads or generations at Buses i and j. Within each iteration, the following equations hold:

$$0 = P_{gi} + P_{is} - P_{di} - V_i \sum_{j \in N_i} V_j (G_{ij} \cos \theta_{ij} + B_{ij} \sin \theta_{ij}) \quad i \in N_{B-1}$$

$$0 = Q_{gi} + Q_{is} - Q_{di} - V_i \sum_{j \in N_i} V_j (G_{ij} \sin \theta_{ij} - B_{ij} \cos \theta_{ij}) \quad i \in N_{PQ}$$

(4)

where

N_i number of buses adjacent to bus i, including bus i

N_{B-1} total numbers of buses, excluding the slack bus

N_{PQ} number of PQ buses

P_{is} and Q_{is} are computed after each iteration but are not used in deriving the Jacobian matrix, so the symmetry property of the bus admittance matrix is maintained and the PQ decoupled power flow can be used without any modification.

9.2.2 Series Compensator

The series compensator is shown in Figure 9.3 as a controllable capacitor. The capacitor is just used simply as a changeable reactance to be implemented in the bus susceptance matrices.

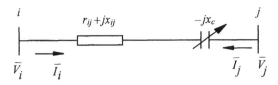

Figure 9.3 Series compensator line

9.3   PROBLEM FORMULATION

The power flow controllers (pfc) are used as phase shifters and series compensators to regulate the power flow. Since the active power loss is used as the objective function and the active power flow limits of the transmission lines during the contingency states are used as security constraints, the objective function of the OPF in FACTS is therefore expressed as follows:

$$\min P_{\text{loss}} = \sum_{\substack{k \in N_{\text{E}} \\ k=(i,j)}} g_k \left(V_i^2 + V_j^2 - 2V_i V_j \begin{cases} \cos\theta_{ij} & \text{if } k \notin N_{\text{pfc}} \\ \cos(\theta_{ij} + \Phi_k) & \text{if } k \in N_{\text{pfc}} \end{cases} \right)$$

$$\text{s.t. } 0 = P_i - V_i \sum_{j \in N_i} V_j (G_{ij} \cos\theta_{ij} + B_{ij} \sin\theta_{ij}) \quad i \in N_{\text{B}-1}$$

$$0 = Q_i - V_i \sum_{j \in N_i} V_j (G_{ij} \sin\theta_{ij} - B_{ij} \cos\theta_{ij}) \quad i \in N_{PQ}$$

$$\tag{5}$$

$$T_k^{\min} \le T_k \le T_k^{\max} \qquad k \in N_{\text{phs}}$$

$$\Phi_k^{\min} \le \Phi_k \le \Phi_k^{\max} \qquad k \in N_{\text{phs}}$$

$$0 \le x_{ck} \le x_{ck}^{\max} \qquad k \in N_{\text{sc}}$$

$$|P_k| \le P_k^{\max} \qquad k \in N_{\text{Elim}}$$

where

N_{E} number of branches

N_{phs} number of phase shifter branches

N_{sc} number of series compensator branches

N_{Elim} number of limited power flow branches

P_{loss} network real power loss

g_{k} conductance of branch k

Power flow equations are used as equality constraints; phase shifter angles, tap settings, series compensator settings and the active power flow of transmission lines are used as inequality constraints. The phase shifter angles, tap settings and series compensator settings are control variables so they are self-restricted. The active power flows of transmission lines are state variables, which are restricted by adding them as the quadratic penalty terms to the objective function to form a penalty function. Equation (5) is therefore changed to the following generalised objective function:

$$\min f = P_{\text{loss}} + \sum_{k \in N_{\text{Elim}}} \lambda_k (|P_k| - P_k^{\max})^2$$

$$\text{s.t. } 0 = P_i - V_i \sum_{j \in N_i} V_j (G_{ij} \cos\theta_{ij} + B_{ij} \sin\theta_{ij}) \quad i \in N_{\text{B}-1} \tag{6}$$

$$0 = Q_i - V_i \sum_{j \in N_i} V_j (G_{ij} \sin\theta_{ij} - B_{ij} \cos\theta_{ij}) \quad i \in N_{PQ}$$

where λ_k is the penalty factor that can be increased in the optimisation process.

9.4 EVOLUTIONARY PROGRAMMING (EP)

EP is different from conventional optimisation methods. It does not need to differentiate cost function and constraints. It uses probability transition rules to select generations. Each individual competes with some other individuals in a combined population of the old generation and the mutated old generation. The competition results are valued using a probabilistic rule. Winners are selected to constitute the next generation and the number of winners is the same as that of the individuals in the old generation.

The procedure of EP for OPF in FACTS is briefly as follows.

9.4.1 Initialisation

The initial control variable population is selected by randomly selecting $p_i = [x_c{}^i,$ $T^i, \Phi^i]$, $i = 1, 2, ..., m$, where m is the population size, from the sets of uniform distribution ranging over $[0, x_c{}^{max}]$, $[T^{min}, T^{max}]$ and $[\Phi^{min}, \Phi^{max}]$. The fitness score f_i of each p_i is obtained by running PQ decoupled power flow.

The rest of the EP procedures are the same as before.

9.5 NUMERICAL RESULTS

In this section, a modified IEEE 30-bus system is used to show the effectiveness of the algorithm. The network is shown in Figure 9.4 and the impedances, loads and power generations, except the generation from the slack bus; that is, Bus 1, are given in [5]. All voltage and power quantities are per-unit values. The total loads are: $P_{load} = 2.834$, $Q_{load} = 1.262$. The voltages at the generator buses, PV buses and a slack bus, are set to 1.0. Four branches, (6,10), (4,12), (10,22) and (28,27) are installed with phase shifters. Three branches, (1,3), (3,4) and (2,5) are installed with series compensators. The phase shifter limits are given as $-20° \leq \Phi_k \leq 20°$ and the series compensator limits are listed in Table 9.1.

Three cases are studied. Case 1 is the normal operation state. Cases 2 and 3 are contingency states. Case 2 has one circuit outage of branch (6,28). Case 3 has two circuit outages of branches (6,28) and (10,21). Two branches, (10,22) and (8,28), are selected as limited power flow branches whose limits are set to 0.1.

The initial phase shifter angles are 0° and the compensator reactances are 0. The optimal phase angles and series compensator reactances are listed in Table 9.2 and generations and power losses are listed in Table 9.3. The power flows of branches (10,22) and (8,28) are regulated back into their limits in both Cases 2 and 3 as in Table 9.4. After the optimisation, there is a 0.00066 per-unit power saving in Case 1. In the other two cases, the active power losses are much more. The regulations of phase angles are used to remove the overloads of branches after the circuit outages.

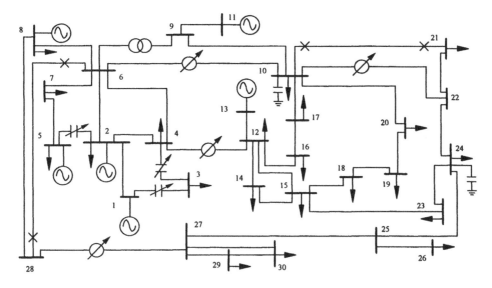

Figure 9.4 The modified IEEE 30-bus system

Table 9.1 Series compensator limits

Branch	(1,3)	(3,4)	(2,5)
$x_C{}^{max}$	0.1	0.02	0.1
$x_C{}^{max}/x_{ij}$ (%)	54	53	50

Table 9.2 Optimal phase angles and compensators

Branch	(6,10)		(4,12)		(10,22)		(28,27)		(1,3)	(3,4)	(2,5)
	Φ^o	T	Φ^o	T	Φ^o	T	Φ^o	T	x_C	x_C	x_C
Case 1	0.52	1.00	-1.11	1.00	-0.04	1.00	0.88	1.03	0.021	0.009	0.021
Case 2	0.03	1.04	-1.72	1.02	-0.10	1.00	-4.93	1.00	0.014	0.02	0.022
Case 3	-0.12	1.10	2.87	1.10	-10.33	0.92	-10.89	0.97	0.038	0.004	0.026

Table 9.3 Generations and power losses

	Initial Conditions				Optimal Results			
	P_g	Q_g	P_{loss}	Q_{loss}	P_g	Q_g	P_{loss}	Q_{loss}
Case 1	2.89388	0.98020	0.05988	-0.28180	2.89321	0.96974	0.05922	-0.29226
Case 2	2.89666	1.00151	0.06267	-0.26049	2.89849	0.99431	0.06450	-0.26769
Case 3	2.90174	1.01477	0.06774	-0.24722	2.92511	1.07865	0.09111	-0.18335

Table 9.4 Power flows of branches

	Initial Conditions		Optimal Results	
Branches	(10,22)	(8,28)	(10,22)	(8,28)
Case 1	0.080	0.021	0.0766	0.0254
Case 2	0.086	0.147	0.100	0.100
Case 3	0.231	0.158	0.100	0.100

9.6   CONCLUSIONS

The potential of the application of EP to the OPF of FACTS has been shown in this chapter. With no need to differentiate the objective function and the constraint equations, the application of EP to unified control strategy can effectively find a solution.

9.7   ACKNOWLEGMENTS

The author wishes to acknowledge the IEEE for granting permission to reproduce the material and results contained in [6].

9.8   REFERENCES

[1]   G N Taranto, L M V G Pinto and M V F Pereira, 'Representation of FACTS devices in power system economic dispatch', *IEEE Transactions on Power Systems*, **7**, 1992, 572-576.

[2]   Z X Han, 'Phase shifter and power flow control', *IEEE Transactions on Power Apparatus and Systems*, **101**, 1982, 3790-3795.

[3]   M Noroozian and G Andersson, 'Power flow control by use of controllable series components', *IEEE Transactions on Power Delivery*, **8**, 1993, 1420-1429.

[4] L L Lai and J T Ma, 'Optimal power flow in FACTS using genetic algorithms', *IEEE/KTH Stockholm Power Tech International Symposium on Electric Power Engineering*, 1995, 484-489.

[5] K Y Lee, Y M Park and J L Ortiz, 'A united approach to optimal real and reactive power dispatch', *IEEE Transactions Power Apparatus and Systems*, **104**, 1985, 1147-1153.

[6] L L Lai and J T Ma, 'Power flow control in FACTS using Evolutionary Programming', *1995 IEEE International Conference on Evolutionary Computation*, 1995.

10

Multi-Time-Interval Scheduling for Daily Operation of a Two-Co-generation System with Evolutionary Programming

This chapter presents a multi-time-interval scheduling for the daily operation of a two-co-generation system connected with auxiliary devices, which include auxiliary boilers, heat storage tanks, electricity chargers and independent generators. The efficiency of a co-generation system depends on the production of thermal and electrical energy that is modelled with a quadratic equation obtained from the least squares method. Evolutionary programming (EP) is used to establish the scheduling of the operation for the co-generation system. For this complex scheduling problem with multivariables in multi-time-intervals, the optimal operation cost for the co-generation system obtained with EP is much lower than the initial feasible solution obtained by trial and error.

10.1 INTRODUCTION

Since the world oil crisis, co-generation systems have been constructed or operated in many countries. A co-generation system can supply both thermal and electrical energies simultaneously from an energy source. Therefore it is known as *total energy system*. It can be constructed easily in urban areas. It has therefore been used in the past as the distributed electric energy source in electricity utilities. Its main advantage is the effective use of input energy to power plants. A co-generation system must be operated with an acceptable scheduling for energy saving.

Many papers related to co-generation systems have been published. [1,2] report the modelling of co-generation systems; [3-5] report the economic dispatch and planning for such systems; and [5-7] present results on the co-generation operation scheduling connected with auxiliary devices. However, little work has been done on the optimal operation scheduling of co-generation systems with heat storage tanks or auxiliary boilers. This is an increasingly important area, which has been included in this chapter.

To operate co-generation systems more effectively, an efficient operation strategy has to be developed. This chapter deals with the operation scheduling problem of an industrial co-generation system operated in the bottoming cycle, which mainly produces thermal energy to supply the thermal load demand, while electric energy production depends on the production of thermal energy. It is one of two types of co-generation system. The other is the topping cycle, which mainly generates electricity but uses the remaining heat and auxiliary boiler to supply some thermal load demand. A co-generation system can be operated by connecting with several auxiliary devices. Therefore, this chapter presents an operation scheduling method for a co-generation system that has four kinds of auxiliary devices such as an auxiliary boiler, heat storage tank, electricity charger and independent generator. The objective of scheduling is to minimise the total operation cost while satisfying system constraints.

The efficiency of co-generation systems depends on the production of heat and electricity. In this chapter the efficiency equation is obtained by the least square method from the data taken over a period of six months. A quadratic polynomial is used to model the efficiency. The objective function has a non-linear characteristic.

Conventional optimisation methods are based on successive linearisations and use the first and second differentiations of the objective function and its constraint equations as the search directions. Conventional optimisation methods are good enough for the optimisation problems of the deterministic quadratic objective function, which has only one minimum. However, the co-generation problem involves both linear and quadratic functions that induce many local minima. Conventional optimisation methods can only lead to a local minimum and sometimes result in divergence. This chapter presents an application of evolutionary programming (EP) to solve the co-generation problem. EP is an artificial intelligence method that is an optimisation algorithm based on the mechanics of natural selection—mutation, competition and evolution. The process of evolution inevitably leads to the optimisation of 'behaviour' within the context of a given criterion.

The simulation is performed for the bottoming cycle co-generation system, which mainly supplies mainly thermal power but uses the remaining heat to produce electric power. The system consists of one or two co-generation units and several different auxiliary devices. This chapter gives a detailed model for practical co-generation units and auxiliary devices. The operational performance of the auxiliary devices is carefully studied. All continuous and discrete variables and functions are easily considered by the EP, while, with conventional optimisation methods, such discrete variables would not be so easy to deal with.

A daily operation scheduling is established based on the load level and characteristics of the co-generation system and auxiliary devices. Different prices for buying and selling the electricity between the co-generation system and electric utility have been considered. The EP approach could achieve a better operation for the co-generation system to minimise the total operation cost. The results show a high potential in using EP for the optimal scheduling of practical industrial co-generation systems.

10.2 MODELLING OF MULTI-CO-GENERATION SYSTEMS

List of symbols

H_{coi} heat output from cogeneration system (MW)

H_{load} heat sending to thermal load (MW)

H_{sti} heat transferring to heat storage tank (MW)

H_{abi} heat output from auxiliary boiler (MW)

H_{wast} waste heat (MW)

H_{rtn} return heat (MW)

E_{coi} electricity output from co - generation system (MW)

E_{bs} electric power buying from or selling to electricity utility (MW)

E_{load} electric power sending to load (MW)

E_{chgi} electricity charging to battery (MW)

E_{indgi} electricity outputs from independent generator (MW)

N total number of time intervals

γ transmission efficiency between the electricity utility and the co - generation plant

η_{ci} efficiency of co - generation system

η_{ai} efficiency of auxiliary boiler

η_{gi} efficiency of independent generator

F_{ci} fuel cost of co - generation system ($ / MWh)

F_{ai} fuel cost of auxiliary boiler ($ / MWh)

F_{gi} fuel cost of independent generator ($ / MWh)

R_{sehi} maximum ratio of electricity to steam of co - generation system

R_{seli} minimum ratio of electricity to steam of co - generation system

E_{stori} electric energy stored in battery (MWh)

E_{lossi} electricity loss coefficient of battery (MWh / h)

H_{stori} heat energy stored in heat storage tank (MWh)

H_{lossi} heat loss coefficient of heat storage tank (MWh / h)

η_{wast} waste heat constant

η_{rtn} return heat constant

C_e price of electricity buying from or selling to the electricity utility ($ / MWh)

μ_{ei} coefficient for electric energy loss

μ_{ti} coefficient for thermal energy loss

t length of each time interval (h)

η_{chgi} charge efficiency of battery

$\eta_{dischgi}$ discharge efficiency of battery

δ_{abi} heat transfer efficiency of auxiliary boiler

δ_{hsti} heat transfer efficiency of heat storage tank

The co-generation plant has two different units and four kinds of auxiliary devices. The energy flow of the proposed multi-co-generation system is illustrated in Figure 10.1. Electricity and heat flow through the electricity bus and heat bus are simplified as a single bus.

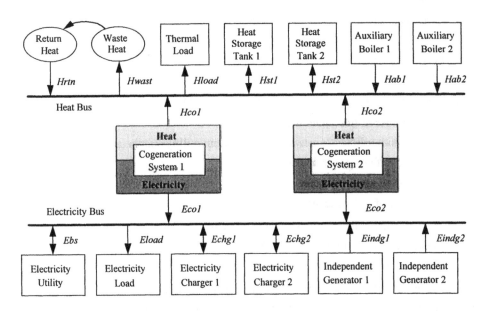

Figure 10.1 Energy flow of multi-co-generation systems

In this chapter, the thermal and electrical loads during the operation period are given in the information from the industrial plants. The efficiencies of independent generators and auxiliary boilers are assumed to be constant. The electrical transmission loss generated between the electricity utility and the industrial plant is charged to the owner of the co-generation plants. On the basis of the above assumptions, the objective function used to minimise the total operation is formulated. The total operation cost is the sum of the fuel cost of the co-generation system, auxiliary boilers and independent generators, and the buying and selling cost of electricity between the co-generation system and the electricity utility.

The objective function is represented by Equation (1). The first term represents the buying and selling cost of electricity while the second term is the total fuel cost of the co-generation system, auxiliary boilers and independent generators.

$$\min \sum_{k=1}^{N} t \left(C_e(k) \begin{cases} \dfrac{1}{\gamma} E_{bs}(k) & \text{if } E_{bs}(k) > 0 \\ \gamma E_{bs}(k) & \text{otherwise} \end{cases} \right.$$

$$+ \sum_{i=1}^{2} \frac{F_{ci}}{\eta_{ci}} H_{coi}(k) + \sum_{i=1}^{2} \frac{F_{ai}}{\eta_{ai}} H_{abi}(k) + \sum_{i=1}^{2} \frac{F_{gi}}{\eta_{gi}} E_{indgi}(k)) \Big)$$

$$\text{s.t. } H_{load}(k) = \left(\sum_{i=1}^{2} \left(H_{coi}(k) + H_{abi}(k) - H_{sti}(k) \right) \right) - H_{wast}(k) + H_{rtn}(k)$$

$$E_{load}(k) = \left(\sum_{i=1}^{2} \left(E_{coi}(k) + E_{indgi}(k) - E_{chgi}(k) \right) \right) + E_{bs}(k)$$

(1)

$$H_{coi}^{min} \le H_{coi}(k) \le H_{coi}^{max}$$
$$R_{seli} H_{coi}(k) \le E_{coi}(k) \le R_{sehi} H_{coi}(k)$$
$$\delta_{abi} H_{abi}^{min} \le H_{abi}(k) \le \delta_{abi} H_{abi}^{max}$$
$$E_{indgi}^{min} \le E_{indgi}(k) \le E_{indgi}^{max}$$

where the efficiencies of co-generation systems 1 and 2 vary according to the production of thermal and electric energy. Each efficiency is obtained as a quadratic equation by the least squares method as below:

$$\eta_{ci}(k) = \frac{1}{100}(a_i + b_i E_{coi} + c_i E_{coi}^2 + d_i H_{coi} + e_i H_{coi}^2)$$

(2)

The electricity is charged according to the following equation:

$$E_{stori}(k+1) = t \begin{cases} \eta_{chgi} E_{chgi}(k) & E_{chgi} > 0 \\ \dfrac{1}{\eta_{dischgi}} E_{chgi}(k) & \text{otherwise} \end{cases}$$

$$+ E_{stori}(k)(1-\mu_{ei})^t \tag{3}$$

$$E_{stori}(0) = E_{stori0}$$

The electric energy stored in the battery is restricted according to the following equation:

$$E_{stori}^{min} \le E_{stori}(k) \le E_{stori}^{max} \tag{4}$$

The thermal energy transferred to the heat storage tank is represented by the following equation:

$$H_{stori}(k+1) = t \begin{cases} \delta_{hsti} H_{sti}(k) & H_{sti}(k) > 0 \\ \dfrac{1}{\delta_{hsti}} H_{sti}(k) & \text{otherwise} \end{cases}$$

$$+ H_{stori}(k)(1-\mu_{ti})^t \tag{5}$$

The thermal energy stored in the heat storage tank is limited between the upper and lower boundaries and is shown in (6):

$$H_{stori}^{min} \le H_{stori}(k) \le H_{stori}^{max} \tag{6}$$

A part of the thermal energy produced in the co-generation system dissipates as the thermal loss, while the other part of the thermal energy, used as process heat, returns to the co-generation system. Therefore, the waste heat and return heat can be represented by the following equation:

$$H_{wast}(k) = \eta_{wast} \sum_{i=1}^{2} \left(H_{load}(k) + H_{sti}(k) - H_{abi}(k) \right)$$

$$H_{rtn}(k) = \eta_{rtn} \sum_{i=1}^{2} \left(H_{load}(k) + H_{sti}(k) - H_{abi}(k) \right) \tag{7}$$

10.3 EVOLUTIONARY PROGRAMMING (EP)

EP is different from conventional optimisation methods. It does not need to differentiate cost function and constraints. It uses probability transition rules to select generations. Each individual competes with some other individuals in a combined population of the old generation and the mutated old generation. The competition results are valued using a probabilistic rule. The same number of winners as the individuals in the old generation constitute the next generation.

The procedure of EP for co-generation scheduling is briefly as follows.

10.3.1 Initialisation

The initial control variable population is selected by randomly selecting $p_i=[H_{co1}(k)^i$, $H_{co2}(k)^i$, $H_{ab1}(k)^i$, $H_{ab2}(k)^i$, $H_{st1}(k)^i$, $H_{st2}(k)^i$, $E_{co1}(k)^i$, $E_{co2}(k)^i$, $E_{indg1}(k)^i$, $E_{indg2}(k)^i$, $E_{bs}(k)^i$, $E_{chg1}(k)^i$, $E_{chg2}(k)^i]$, $i=1, 2, ..., m$, where m is the population size, from the sets of uniform distribution. The fitness score f_i of each p_i is obtained by calculating the objective function and taking Equations (2) to (7) into account. The rest of the EP procedures are the same as before.

10.4 CASE STUDY

The simulated co-generation plant is shown schematically in Figure 10.1. There are two co-generation systems, two auxiliary boilers, two heat storage tanks, two independent generators and two electricity chargers in the plant. The plant supplies both thermal and electric loads and is connected to the electricity utility. An eight-time interval scheduling is used for the daily operation of the system, with a three-hour period in each interval. The co-generation system parameters are listed in Table 10.1. The daily loads and electricity buying-selling prices are given in Table 10.2. The coefficients of the efficiency equation for the co-generation systems given in Table 10.3 are obtained by the least squares method on the basis of the measurement data from a practical co-generation plant. The electricity buying-selling prices are provided.

Table 10.1 Co-generation system parameters

E_{stor1}^{max}	5.0	H_{stor10}	1.0	H_{col}^{max}	25.0	μ_{T1}	0.05	δ_{ab1}	0.98
E_{stor1}^{min}	1.0	H_{stor20}	1.0	H_{col}^{min}	5.0	μ_{T2}	0.05	δ_{ab2}	0.98
E_{stor2}^{max}	6.0	H_{ab1}^{max}	5.0	H_{co2}^{max}	30.0	μ_{E1}	0.05	δ_{hst1}	0.98
E_{stor2}^{min}	1.0	H_{ab1}^{min}	1.0	H_{co2}^{min}	7.0	μ_{E2}	0.05	δ_{hst2}	0.98
E_{stor10}	1.0	H_{ab2}^{max}	7.0	F_{a1}	19.55	η_{a1}	0.7	R_{seh1}	0.85
E_{stor20}	1.0	H_{ab2}^{min}	1.0	F_{a2}	19.55	η_{a2}	0.7	R_{sel1}	0.65
H_{stor1}^{max}	6.0	E_{indg1}^{max}	5.0	F_{c1}	19.55	η_{g1}	0.5	R_{seh2}	0.9
H_{stor1}^{min}	7.0	E_{indg1}^{min}	1.0	F_{c2}	19.55	η_{g2}	0.5	R_{sel2}	0.7
H_{stor2}^{max}	1.0	E_{indg2}^{max}	7.0	F_{g1}	25.75	η_{wast}	0.3	t	6.0
H_{stor2}^{min}	1.0	E_{indg2}^{min}	1.0	F_{g2}	25.75	η_{rtn}	0.15	γ	0.98

Table 10.2 Loads and electricity buying-selling prices

k	1	2	3	4	5	6	7	8
$H_{load}(k)$	53.06	53.06	55.22	55.22	60.33	60.33	53.06	53.06
$E_{load}(k)$	50	50	64	64	70	70	65	65
$C_e(k)$		31.54	31.54	82.44	82.44	82.44	82.44	58.21	58.21

Table 10.3 Coefficients for co-generation system efficiency equation

i	a_i	b_i	c_i	d_i	e_i
1	50.0	0.0083	-0.00000029	0.90	-0.0032
2	49.0	0.0085	-0.00000031	0.85	-0.0038

10.4.1 Initial Result

The initial feasible variable settings are given in Tables 10.4 and 10.5. The units for all variables are MW, except for $H_{stor1}(k)$, $H_{stor2}(k)$, $E_{stor1}(k)$ and $E_{stor2}(k)$, whose units are MWh. The production cost for the initial setting is $69103.5.

Table 10.4 Initial thermal power results

k	$H_{co1}(k)$	$H_{co2}(k)$	$H_{ab1}(k)$	$H_{ab2}(k)$	$H_{st1}(k)$	$H_{stor1}(k)$	$H_{st2}(k)$	$H_{stor2}(k)$	$H_{wast}(k)$	$H_{rtn}(k)$
1	25	25	4.9	6.86	1.7492	6.0	0.4291	2.1188	13.0435	6.5217
2	25	24	4.9	6.86	0.2911	6.0	1.0176	4.8084	12.7826	6.3913
3	25	26	4.9	6.86	0.2911	6.0	0.5968	5.8771	13.3043	6.6522
4	25	26	4.9	6.86	0.2911	6.0	0.5968	6.7933	13.3043	6.6522
5	25	30	4.9	6.86	-0.7439	2.8670	0.0	5.8244	14.3478	7.1739
6	25	30	4.9	6.86	-0.4763	1.0	-0.2676	4.1745	14.3478	7.1739
7	25	24	4.9	6.86	1.3087	4.7049	0.0	3.5791	12.7826	6.3913
8	25	24	4.9	6.86	0.6687	6.0	0.64	4.9501	12.7826	6.3913

Table 10.5 Initial electrical power results

k	$E_{co1}(k)$	$E_{co2}(k)$	$E_{indg1}(k)$	$E_{indg2}(k)$	$E_{bs}(k)$	$E_{chg1}(k)$	$E_{stor1}(k)$	$E_{chg2}(k)$	$E_{stor2}(k)$
1	21.25	22.5	5	7	-3	1.5343	5.0	1.2157	4.1398
2	21.25	21.6	5	7	-4	0.2641	5.0	0.5859	5.1312
3	21.25	23.4	5	7	8	0.2641	5.0	0.3859	5.4412
4	21.25	23.4	5	7	8	0.2641	5.0	0.3859	5.7071
5	21.25	27	5	7	10	0.25	4.9619	0.0	4.8931
6	21.25	27	5	7	10	0.25	4.9292	0.0	4.1952
7	21.25	21.6	5	7	10	-0.15	3.7262	0.0	3.5969
8	21.25	21.6	5	7	10	-0.15	2.6947	0.0	3.0839

10.4.2 Optimal Result

After a number of EP runs, the optimal result is given in Tables 10.6 and 10.7. The production cost is $57433.6. The saving is:

$$\text{Saving} = 69103.5 - 57433.6 = \$11669.9$$

$$\text{Saving\%} = \frac{11669.9}{69103.5} \times 100 = 16.89\%$$

With the EP approach, it can be seen that a very attractive saving for the multi-time-interval scheduling for the daily operation of the two co-generation systems and multiauxiliary devices has been achieved. The CPU time is 4 min and 37 s when a 166 MHz Pentium is used.

Table 10.6 Optimal thermal power results

k	$H_{co1}(k)$	$H_{co2}(k)$	$H_{ab1}(k)$	$H_{ab2}(k)$	$H_{st1}(k)$	$H_{stor1}(k)$	$H_{st2}(k)$	$H_{stor2}(k)$	$H_{wast}(k)$	$H_{rtn}(k)$
1	24.7837	29.0768	3.99334	5.95597	1.7492	6.0	1.9753	6.6647	14.0506	7.0253
2	24.6142	29.7146	0	6.64066	-1.2169	1.4190	0.0	5.7142	14.1727	7.0864
3	24.9923	29.9609	4.80617	6.83537	0.2469	1.9426	0.0	4.8992	14.3356	7.1678
4	24.9684	29.991	4.7538	6.53771	-0.2174	1.0	-0.9204	1.3830	14.3372	7.1686
5	25	29.9164	4.79726	6.84242	0.1731	1.3663	0.0	1.1857	14.3260	7.1630
6	25	29.9952	4.9	5.38455	0.9064	3.8364	0.0	1.0166	14.3466	7.1733
7	24.5807	29.5556	4.00592	6.8463	0.9220	6.0	1.9053	6.4731	14.1225	7.0613
8	24.9386	30	0	2.61187	-1.3538	1.0	-1.3217	1.5040	14.3318	7.1659

Table 10.7 Optimal electrical power results

k	$E_{co1}(k)$	$E_{co2}(k)$	$E_{indg1}(k)$	$E_{indg2}(k)$	$E_{bs}(k)$	$E_{chg1}(k)$	$E_{stor1}(k)$	$E_{chg2}(k)$	$E_{stor2}(k)$
1	21.0661	26.1691	2.09255	0	-2.68484	1.5343	5.0	0.1086	1.1507
2	20.922	26.7431	0	2.74352	6.36407	0.2641	5.0	1.5086	5.0599
3	21.2435	26.9648	2.60774	0	14.8207	0.2641	5.0	0.3726	5.3441
4	21.2231	26.9919	0	6.98167	14.317	-0.4863	2.6659	0.0	4.5819
5	21.25	26.9248	4.43556	6.38004	9.12483	1.0053	5.0	0.1099	4.2252
6	21.25	26.9957	0	6.43859	3.64887	-0.4013	1.0	-0.6808	1.3532
7	20.8936	26.6	4.98334	5.66294	-2.72657	0.7989	5.0	0.0	1.1602
8	21.1978	27	0	0	1.64052	-0.4013	1.0	0.0	0.9947

10.5 CONCLUSIONS

This chapter has presented the application of EP to multi-time-interval scheduling for co-generation system operation. EP could easily solve such complex scheduling problem with multi-variables in the multi-time-intervals. The simulation results show a high potential for the proposed co-generation model and the operation scheduling method to be applied to the industrial multi-co-generation system, which usually has several c-ogeneration units and multi-auxiliary devices.

10.6 ACKNOWLEGMENTS

The author wishes to acknowledge Elsevier for granting permission to reproduce the material and results contained in [8].

10.7 REFERENCES

[1] H B Püttgen and P R MacGregor, 'Optimum scheduling procedure for cogenerating small power producing facilities', *IEEE Transactions on Power Systems*, **4**, 1989, 957-964.

[2] H Ghoudjehbakloum and H B Püttgen, 'Optimisation topics related to small power producing facilities: operating under energy spot pricing polices', *IEEE Transactions on Power Systems*, **2**, 1987, 296-302.

[3] F J Rooijers and R A M Van Amerongen, 'Static economic dispatch for cogeneration system', *IEEE Transactions on Power Systems*, **9**, 1994, 1392-1398.

[4] S M Wang, C C Liu and S Luu, 'A negotiation methodology and its application to cogeneration planning', *IEEE Transactions Power Systems*, **9**, 1994, 869-875.

[5] K Moslehi, M Khadem, R Berrial and G Hernandez, 'Optimization of mulitplant cogeneration system operation including electric and steam networks', *IEEE Transactions Power Systems*, **6**, 1991, 484-490.

[6] M L Baughman, N Eisner and P S Merrill, 'Optimizing combined cogeneration and thermal storage systems: an engineering economics approach', *IEEE Transactions on Power Systems*, **4**, 1989, 974-980.

[7] J B Lee, S H Lyu and J H Kim, 'A daily operation scheduling on cogeneration system with thermal storage tank', *Transactions of IEE of Japan*, **114-B**, 1994, 1295-1302.

[8] L L Lai, J T Ma and J B Lee, 'Multi-time interval scheduling for daily operation of a two-cogeneration system with evolutionary programming', *International Journal of Electrical & Energy Systems*, **20**, 1998, 305-311.

11

Application of Evolutionary Programming to Fault Section Estimation

This chapter presents an application of evolutionary programming (EP) to fault section estimation in power systems. Several techniques have been employed to solve this problem so far. The genetic algorithm(GA) has been reported to be one of these techniques. To measure the efficiency of EP, the GA is also used to solve the same problem. Different parameters that affect the EP convergence are investigated. Software has been developed to implement both the EP and GA and applied to a sample power system. It shows that EP is superior to GA for the type of coding strategy and evolution used.

11.1 INTRODUCTION

To enhance service reliability and to reduce power outage, rapid restoration of a power system is required. As a first step of restoration, the fault section should be estimated quickly and accurately. The fault section estimation (FSE) identifies fault components in a power system by using information on the operation of protective relays and circuit breakers. However, this task is difficult, especially for cases where the relay or circuit breaker fails to operate and for multiple faults. Several papers have reported surveys on evolutionary algorithm (EA) applications in power systems [1,2]. Few methods have been employed so far to solve the FSE problem. These include, expert systems [3] and other computational intelligence techniques (CIT), such as, artificial neural networks [4,5] and genetic algorithms [6]. Among these methods, expert systems are based on production rules and involve a great number of rules describing complex protection system behaviour. This results in the following problems:

1. The maintenance of such a complex rule-based knowledge base is very difficult and time consuming.

2. The response time for complex systems is long.

The application of ANNs to FSE is an active research area. However, the correctness of the estimation result cannot be proved theoretically. As a result the system is questionable. As the FSE objective function is usually a high-order polynomial, the GA optimisation method has been employed to deal with such a problem [6]. Evolutionary programming is an optimisation algorithm based on the mechanics of natural selection, mutation, competition and evolution. The process of evolution inevitably leads to the optimisation of 'behaviour' within the context of a given criterion. EP does not require a crossover operation and it has a shorter run time when compared with GA [7].

This is the first time EP has been used for FSE. This chapter is presented in the following sections. After formulating the problem in Section 11.2, the derivation of the fitness function and EP parameter is given in Section 11.3. The EP algorithm is described in Section 11.4 and a case study for a modelled power system is introduced and examined by the developed algorithm in section 11.5. The conclusion is given in Section 11.6.

A C++ code has been developed to implement the proposed EP and GA. A Visual C++ compiler is used on a Windows NT platform running on a Pentium 166 machine.

11.2 PROBLEM FORMULATION

To simplify the explanation of the proposed approach, a model is developed. It is assumed that the modelled power system consists of the following components and parameters:

- Circuit Breaker (CB),
- Relay (R),
- Busbar (BB),
- Transformer (T),
- Transmission Line (L), and
- Section i (A_i)

There are three types of relays:

- Main Protective Relay(MPR),
- Primary Backup Protective Relay (PBPR), and
- Secondary Backup Protective Relay (SBPR).

The operation and description of the protective relays are as follows.

1. Each **Busbar** has one MPR. It is used to initiate the circuit breakers to disconnect the fault in the busbar.

2. Each **Transformer** has three relays namely, MPR, PBPR and SBPR. The **MPR** is used to initiate the two circuit breakers at its ends. The **PBPR** is used to initiate the circuit breakers when a fault occurs on one of its neighbouring elements while the main protective relay fails to operate. The purpose of the **SBPR** is to protect the transformer in the case of a fault occurring on one of its neighbouring elements and the main protective relay of the faulted element failing to operate.

3. Each **Transmission Line** has two sets of MPR, PBPR and SBPR; one for the sending end and one for the receiving end. The **MPR** at each end is used to actuate the circuit breaker at that end when there is fault on the line. The **PBPR** at each end is used to protect the line in case a fault occurs while the main relay fails to operate. The **SBPR** is used to protect the transmission line in case a fault occurs on one of its neighbouring elements but the main protective relay fails to operate.

11.3 MATHEMATICAL MODEL AND FITNESS FUNCTION

To use the EP technique, the problem has to be modelled mathematically. The EP technique is based on the assumption that a fitness landscape can be characterised in terms of variables and there is one or a set of optimum solutions. A mathematical model has been introduced in [5] and is used to formulate a 0-1 integer programming problem which is then solved by the GA. Although the GA and EP have similar data dictionaries, they have a different data structure. EP has more flexibility for the selection of a data type. It operates directly on an individual type, e.g. a floating-point. A form of the same mathematical model used in [5] is adapted in this chapter. This modified version can be operated in conjunction with EP characteristics in the forms of a [min, max] floating-point programming problem.

The fitness function is one of the main elements of an EP algorithm. The fitness function evaluates each individual and returns a value indicating how fit that individual is, to be considered as a solution to the problem. Each component of the system corresponds to a mathematical term that appropriately affects the final value of the fitness function. The statuses of busbars and relays are presented by two values, 0 for non-operational and 1 for operational conditions. Each section of the power system model is considered as an individual for EP. This individual varies over a pre-defined range, i.e. $A_{min} < A_i < A_{max}$. In this chapter A_{max} is 1 and A_{min} is 0.

The combination of 0-1 integers from relays and busbars statuses and the floating-point value of individuals created by EP results in an appropriate value to show the fitness of that particular individual. The final output of the fitness function is then deducted from a large positive constant number in order to secure a positive fitness.

By considering the above points, the corresponding mathematical terms for each relay are directly related to the protection system explained in Section 11.2 and are listed as follows:

- The MPR of any busbar, transformer and sending and receiving ends of each transmission line:

$$\left(1 - 2\left(A_i\_\text{MAIN}\right)\right)A_i \tag{1}$$

$$\left(1 - 2\left(T_i\_\text{MAIN}\right)\right)T_i \tag{2}$$

$$\left(1 - 2\left(L_iS\_\text{MAIN}\right)\right)L_i \tag{3}$$

$$\left(1 - 2\left(L_iR\_\text{MAIN}\right)\right)L_i \tag{4}$$

where MAIN shows the main protective relay of each component and A_i, T_i, and L_i are the sections.
- The PBPR of any transformers, sending and receiving ends of each transmission line:

$$\left(1 - 2\left(T_i\_\text{PRIM}\right)\right)T_i\left(1 - T_i\_\text{MAIN}\right) \tag{5}$$

$$\left(1 - 2\left(L_iS\_\text{PRIM}\right)\right)L_i\left(1 - L_iS\_\text{MAIN}\right) \tag{6}$$

$$\left(1 - 2\left(L_iR\_\text{PRIM}\right)\right)L_i\left(1 - L_iR\_\text{MAIN}\right) \tag{7}$$

Where PRIM and MAIN show the primary back-up and main protective relays respectively for each component. T_i and L_i are the sections.
- The SBPR of any transformer:

$$\left(1 - 2\left(T_i\_\text{SEC}\right)\right) \times \left\{1 - \left[1 - A_j\left(1 - A_j\_\text{MAIN}\right)\right]\left[1 - A_k\left(1 - A_k\_\text{MAIN}\right)\right]\right\} \times \left(\text{CB}_l | \text{CB}_n\right) \tag{8}$$

Where T_i_SEC is the SBPR of transformer T_i, A_j is one of the sections protected by this relay, A_j_MAIN is the main relay of this section, A_k is the other section that is protected by this relay and A_k_MAIN is its main relay. CB_l and CB_n are neighbouring circuit breakers of the transformer.
- The SBPR of any transmission line:
  For the sending end:

$$\left(1 - 2\left(L_i S\_SEC\right)\right)\left\{1 - \left[1 - A_j\left(CB_k\right)\right]\right\}$$ (9)

For the receiving end:

$$\left(1 - 2\left(L_i R\_SEC\right)\right)\left\{1 - \left[1 - A_j\left(CB_k\right)\right]\right\}$$ (10)

Where $L_i S\_SEC$ and $L_i R\_SEC$ are the secondary sending and receiving end relays, respectively. CB_k is the corresponding circuit breaker of the corresponding section A_j.

Each configuration of circuit breakers and relays produces an input pattern. A list of possible fault sections forms the system output.

11.4 EVOLUTIONARY PROGRAMMING (EP)

EP is a computational intelligence method in which an optimisation algorithm is the main engine for the process of three steps namely, natural selection, mutation and competition. According to the problem, each step could be modified and configured in order to achieve the optimum result. Each possible solution to the problem is called an individual. The mathematical form of the i individual is

$$p_i = \left[A_k^i\right], k = 1,2,\dots,m$$ (11)

where m is the maximum number of parameters in any possible solution and $A_{\min} < A_k^i < A_{\max}$. To use EP, the mathematical model should be capable of dealing with the data type and structure of individuals.

11.4.1 Initialisation

The initial population consists of individuals (sections) and is created randomly. The fitness score f_i of each p_i is obtained by a fitness function. The fitness function is the summation of the corresponding terms of each protective relay and circuit breaker. The rest of the EP procedures are the same as before.

11.5 GENETIC ALGORITHMS (GAS)

Genetic algorithms are the most popular and widely used of all the evolutionary algorithms. They have been widely applied to solve complex non-linear optimisation problems in a number of engineering disciplines. They operate on a population of strings (chromosomes) that encode the parameter set of the problem to be solved over some finite alphabet. In the chosen case study, there are 10 parameters. Each parameter is encoded into a eight bit binary string (gene) producing an 80 (8 x 10) bit chromosome. As the range of each parameter is from 0 to 1, this gives us the step size (precision) of 0.0039, which is quite acceptable for our case. Each encoding represents an individual in the GA population. The population is initialised to random individuals (random chromosomes) at the start of the GA run. The GA searches the space of possible chromosomes (Hamming space) for better individuals. The search is guided by fitness values returned by the modelled protection strategy explained earlier. The same fitness function as that used in EP is adopted. This gives a measure of how well each individual is in terms of solving the problem and hence determining its chance of appearing in future generations. Two-point crossover is used and the crossover probability and mutation probability are 0.85 and 0.1, respectively. The Elitist generation method is used to transfer the best individual through the generations and therefore guarantee the convergence.

11.6   EP VS. GA

Considering both algorithms, it can be concluded that the EP run needs less computational time than the GA run. This has also been proven by the tests using the developed C++ code for each algorithm. The difference in computational speed can be related to the characteristic of each algorithm. GA individuals are represented in a binary form; hence, crossover and mutation are binary operators. Execution of these operators takes a longer time in comparison with the simple EP mutation operator, especially when the number of system-free parameters is high and chromosomes are long and/or the population size is large. The following tests demonstrate the efficiency of each algorithm for the specific problem. In the following section, the results with both EP and GA are presented. The presented graphs, which show the variation of the maximum fitness during the generations, can be used to compare the behaviour of EP and GA.

11.7   CASE STUDY

Figure 11.1 shows a power network that is used to demonstrate the capability of the EP-based or GA-based algorithm. The network consists of 10 sections, 28 relays and 14 circuit breakers.

The sections, protective relays and their assignments are as follows:

A1-A10:   BB1, BB2, BB3, BB4, T1, T2, T3, T4, L1, L2.
R1-R12:   BB1_MAIN,   BB2_MAIN,   BB3_MAIN,   BB4_MAIN,   T1_MAIN, T2_MAIN,   T3_MAIN,   T4_MAIN,   L1R_MAIN,   L1S_MAIN, L2R_MAIN, L2S_MAIN.
R13-R28:   T1_PRIM, T2_PRIM, T3_PRIM, T4_PRIM, T1_SEC, T2_SEC, T3_SEC, T4_SEC, L1R_PRIM, L1S_PRIM, L2R_PRIM, L2S_PRIM, L1R_SEC, L1S_SEC, L2R_SEC, L2S_SEC.

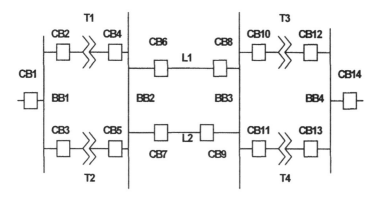

Figure 11.1 Power network

Three different cases have been considered and the fitness has been calculated for each of them. The conditions and results are shown in Tables 11.1 to 11.3 and Figures 11.2 to 11.4.

For Case 1, the operated relays and circuit breakers are given in Table 11.1. The number of generations, population size and estimated fault sections are also presented in the same table. Figure 11.2 shows the average, maximum, minimum and difference between maximum and minimum fitness.

In this case, a possible explanation could be as follows:

BB3_MAIN, L1S_SEC and L2S_SEC fail to trip. CB10 is actuated, so T3_SEC may operate but the signal has not been received by the control centre. CB9 may be tripped by L2R_SEC. CB9 operates but again no signal has been received by the control centre.

For Case 2, the operated relays and circuit breakers are given in Table 11.2. The number of generation, population size and estimated fault sections are also presented in the same table. Figure 11.3 shows the average, maximum, minimum and difference between maximum and minimum fitness.

Table 11.1 The status of relays and circuit breakers and results

Gen. No.	Size of Pop.	Actuated Relays	Actuated Circuit Breakers	Results: Fault Section(s) EP	Results: Fault Section(s) GA
200	10 & 50	L1R_SEC, L2R_SEC, T4_SEC	CB8, CB10, CB11	BB3	BB3

For Case 3, the operated relays and circuit breakers are given in Table 11.3. The number of generation, population size and estimated fault sections are also presented in the same table. Figure 11.4 shows the average, maximum, minimum and difference between maximum and minimum fitness.

This computer simulation on average will give an optimum result in about 60 to 100 generations. The population size is as small as 10, which speeds up the processing time. All the simulation results are performed on a Pentium 166 MHz PC. The CPU time for each case is about 2.5 seconds.

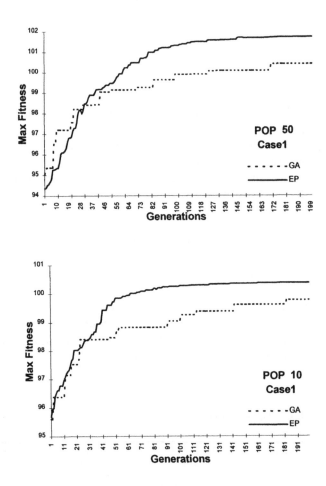

Figure 11.2 Maximum fitness against number of generations, Case 1

Table 11.2 The status of relays and circuit breakers and results

Gen. No.	Size of Pop.	Actuated Relays	Actuated Circuit Breakers	Results: Fault Sections EP	Results: Fault Sections GA
200	10 & 50	T1_SEC, T2_SEC, L1S_MAIN, L1R_MAIN, L1S_PRIM	CB4, CB5, CB8	BB2, L1	BB2, L1

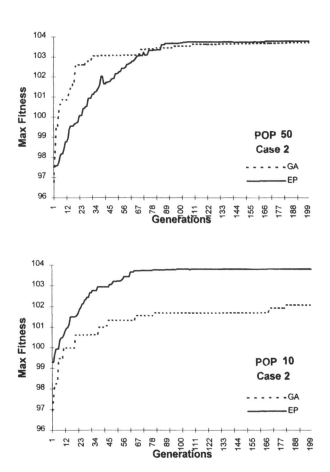

Figure 11.3 Maximum fitness against number of generations, Case 2

Table 11.3 The status of relays and circuit breakers and results

Gen. No.	Size of Pop.	Actuated Relays	Actuated Circuit Breakers	Results: Fault Sections EP	Results: Fault Sections GA
200	10 & 50	T3_SEC, T4_PRIM, BB4_MAIN	CB13, CB14, CB10	BB4, T4	BB4, T4

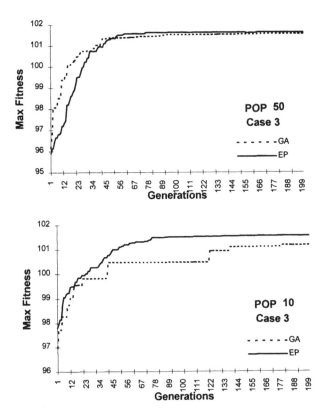

Figure 11.4   Maximum fitness against number of generations, Case 3

11.8 CONCLUSIONS

An EP approach has been developed for solving the FSE problem, including mal-functions of protective relays and/or circuit breakers and multiple fault cases. At the same time, comparison was made with the GA approach. Two different population sizes were tested for each case. In general, EP showed a faster computational speed than GA with an average factor of 13 times greater. The final results were almost the same. The convergence speed (the required number of generations to get an optimum result) is a very important factor in real time applications. Test results show that EP is better than GA for the type of coding strategy and evolution as used for GA. However, as both EP and GA are evolutionary algorithms, their efficiencies are largely dependent upon the complexity of the problem, which might differ from case to case. EP is also ideal for parallel processing computer systems or hardware. Therefore, with this kind of equipment such as transputers, it is possible to solve the FSE problem faster and with high efficiency.

11.9 ACKNOWLEGMENTS

The author wishes to acknowledge the IEE for granting permission to reproduce the material and results contained in [8].

11.10 REFERENCES

[1] D Srinivasan, F Wen, C S Chan and A C Liew, 'A survey of applications of evolutionary computing to power system', *Proceedings of the International Conference on Intelligent Systems Applications to Power Systems*, O A Mohammed and K Tomsovic, Editors, IEEE Catalog Number 96 TH 8152, Jan/Feb 1996, 35-41.

[2] V Miranda, D Srinivasan and L M Proenca, '*Evolutionary computation in power systems'*, *Proceedings of the twelfth Power Systems Computation Conference*, PSCC, Germany, Aug 1996, 25-40.

[3] C Fukui and J Kawakami, 'An expert system for fault section estimation using information from protective relays and circuit breakers', *IEEE Transactions on Power Delivery*, **1**, 1986, 83-90.

[4] L L Lai, F Ndeh-Che, K H Chu, P Rajroop and X F Wang, 'Design of neural networks with genetic algorithms for fault section estimation', Proceedings of the twenty-ninth Universities Power Engineering Conference, **2**, Galway, Ireland, 1994, 596-599.

[5] H T Yang, W Y Chang, and C L Huang, 'A new neural networks approach to on-line fault section estimation using information of protective relays and circuit breakers', *IEEE Transactions on Power Delivery*, **9**, 1994, 220-230.

[6] F Wen and Z Han, 'An evolutionary optimisation method to fault section estimation using information from protective relays and circuit breakers', *Proceedings of the International Conference on Power System Technology*, Beijing, China, International Academic Publishers, 1994, 1051-1055.

[7] J T Ma and L L Lai, 'Evolutionary programming approach to reactive power planning', *IEE Proceedings - Generation, Transmission and Distribution*, **143**, 1996, 365-370.

[8] L L Lai, A G Sichanie and B J Gwyn, 'A comparison between evolutionary programming and genetic algorithm for fault section estimation', to be published in *IEE Proceedings - Generation, Transmission and Distribution*.

12

Neural Networks for Fault Diagnosis in HVDC Systems

This chapter presents the use of neural networks (NNs) to identify faults occurring in a high voltage direct current (HVDC) power transmission system. On the basis of the ability of these NNs to distinguish rapidly and reliably between different types of faults, measures can be taken either to protect the necessary equipment from damage, or to improve the dynamic performance of the a.c.-d.c. power system.

12.1 INTRODUCTION

An important application area for artificial intelligence (AI) techniques is in the fault detection and diagnosis (FDD) of power systems [1-7]. Basically, such techniques operate by mapping fault symptoms (i.e. fundamental frequency phasors of system voltages/currents and related quantities, relay or circuit breaker status, etc.) according to certain algorithms to arrive at diagnostic conclusions. The primary objective of FDD is to limit the damage, repair costs, outage time and danger to the power system equipment. An important secondary objective is to optimise the dynamic performance of the system.

Application of AI-based FDD in HVDC transmission systems with their fast controllers is particularly interesting since the stability of attached a.c. systems can be enhanced if proper control actions are taken when a fault is encountered. This implies an ability to detect rapidly the type of fault encountered, its severity and its location [1]. Consequently, only local fault diagnosis is performed by sensing of the three phase voltages at the converter busbar, the d.c. voltage/current on the d.c. side of the converter, valve firing pulses, etc. Such simple fault detectors sometimes cannot distinguish adequately between the different fault types; as a result, d.c. controller parameters are seldom adapted to provide optimal responses whenever faults are experienced. The causes for false detection may be either due to sensor- or interpretation-related problems. The fault detection sensors are vulnerable to harmonics and non-linearities. The interpretation-related problems may be due to the measurement of fundamental phasors using classical digital algorithms that demonstrate a questionable performance when operating under non-sinusoidal situations [8].

NNs are useful in FDD applications since the nature of the input-output functional relationship is neither well defined nor easily computable. Furthermore, NNs are able to compute the answer quickly by using associations gained either from time-domain simulations or previously gained practical experience. In this chapter, the pattern recognition capabilities of NNs are explored to perform FDD with emphasis on applications in HVDC systems.

12.2 FAULT DIAGNOSIS

A fault is a one-to-many mapping of the fault symptoms. Conversely, fault diagnosis (FD) is a many-to-one mapping, and hence is a much more complex mapping. FD is complicated further since different faults may share common symptoms, and the possibility of multiple faults also exists. In a single fault diagnosis, a high level of correctness is probable. In multiple fault diagnosis, however, this correctness level may be much reduced. FD is not a static problem since various tests may be performed dynamically to identify the fault(s). In addition, faults may evolve over time. Diagnostic problem solving consists of: (a) representation of the problem to be solved, (b) defining strategies for its solution, and finally (c) defining heuristics to guide these strategies.

12.3 PATTERN CLASSIFICATION

Pattern recognition (PR) techniques consist of defining a pattern vector whose components consist of all the significant variables of the primary system variables. To find the most appropriate variables for the primary vector, a thorough understanding of the problem is required. This vector is evaluated in many typical operating conditions to generate a training set. The pattern classification problem is a two-stage process involving feature extraction and classification.

12.3.1 Feature Extraction

The components of the pattern vector are subject to a process of dimensionality reduction, called feature extraction, to achieve three objectives:

- Select the most useful information from the primary vector and represent it in the form of a feature vector of lower dimensionality,
- Remove any redundant/irrelevant information that may have a detrimental effect on the prediction system, and

- Rearrange the variables in terms of their discriminatory effect to provide the consecutive design stages with the most informative variable to be considered.

Statistical methods are used to eliminate redundant information and, as a result, reduce the number of features.

12.3.2 Classification Procedure

This matches input pattern characteristics with a training set to decide upon a classification category. The procedures start from linear combinations of the variable measurements, even though these combinations are subsequently subjected to some non-linear transformation, i.e. MLP is of this type.

The classifiers (Figure 12.1) typically rely on distance metrics and probability theories to perform the above task. In pattern classification, a boundary equation describing the classification (discriminate function) is embedded in the NN by training it with data and an appropriate training algorithm.

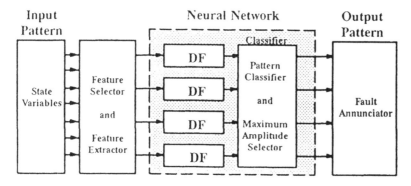

DF = Discriminant Function

Figure 12.1 Fault diagnosis as a pattern classification problem

12.4 HVDC SYSTEMS

A typical HVDC transmission system (Figure 12.2) consists of a bipolar two-terminal scheme. One terminal (rectifier) converts a.c. power to d.c. for transmission over a bipolar d.c. line usually at high voltages to reduce power transmission losses. The other terminal (inverter) reconverts the d.c. power to a.c. for load requirements. HVDC transmission becomes economical over long distances when compared with a.c. transmission owing to greater power transmission per conductor, a simpler line construction and no requirement for charging currents. On the other hand, converters are expensive, require reactive power compensation and need a.c. filters to remove the harmonics generated. Since the d.c. line transmits direct current, which does not have a natural zero-crossing, it is difficult to protect it against d.c. line faults since d.c. breakers are very expensive.

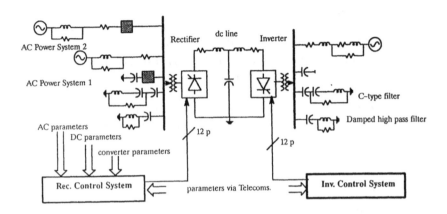

Figure 12.2 Typical HVDC system

A potential FDD application relates to HVDC systems owing to their fast control response (of the order of tens of milliseconds). A two-terminal HVDC system is typically maintained in constant current (CC) control at the rectifier, and in constant extinction angle (CEA) control at the inverter (Figure 12.3).

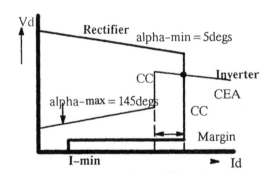

Figure 12.3 Control characteristic

12.4.1 Rectifier Current Controller

A typical HVDC rectifier current controller (Figure 12.4) has two inputs: a limited current order Io lim and the measured d.c. current Id. The two signals are compared and the generated error signal is fed through a classical PI regulator to produce a firing angle order. Appropriate limits to the excursions of an alpha order are usually imposed at alpha-min = 5° and alpha-max = 145°. The alpha order is then fed to a voltage controlled oscillator (VCO) to generate equidistant firing pulses for the converter thyristor valves. The PI regulator has been used historically primarily for its robustness. The problems of choosing the gains of these controllers for an optimal response are well known. Furthermore, the fixed gains of the regulator are optimal only over a narrow operating zone where its performance is considered satisfactory. Utilisation of the FDD techniques to impose gain scheduling for different operating modes of the regulator will improve system performance. Further improvements in control performance may be forthcoming with the use of modern adaptive controllers or intelligent NN [2] and fuzzy logic (FL)-based controllers [9].

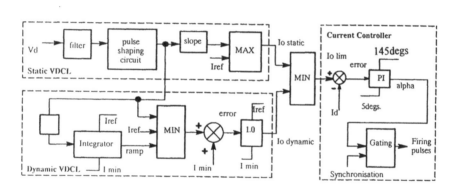

Figure 12.4 Voltage-dependent current (order) limit unit and current controller

Voltage dependent current limits (VDCL) are used in HVDC control systems to limit the converter operation to safe zones. VDCL units have both static/dynamic characteristics (Figure 12.4), and provide the limited current order Io lim to protect the converter valves during system faults and assist the dynamic recovery during the post-fault period. The VDCL unit can be considered as a knowledge-based (KB) pre-processor [8] for the current order of the system controller. The required KB is derived from the following parameters: (a) a.c. system strength and a.c. voltage magnitude; (b) d.c. voltage/current magnitude; (c) d.c. current (or power) demand setting; (d) a.c.-d.c. fault and power quality/harmonic conditions and (e) other parameters, i.e. status of breakers/switchcs (based on system topology).

12.4.2 Inverter CEA Controller

A constant extinction angle (CEA) is desired at the inverter to reduce the incidence of commutation failure (CF), and minimise the converter requirement for reactive power. A CF collapses the d.c. voltage and results in a loss of d.c. power transmission. A typical CEA control scheme uses the error between the desired and actual extinction angles to derive the inverter firing angle via a PI regulator. Again, in common with the rectifier current controller, limited intelligence is used to obtain the required control actions, which are based on 'final effects' rather than on 'primary causes' that may lead to a CF.

One particular challenge for the fast control and protection applications remains in the area of early prediction and detection of a CF at a converter station. Some of the primary causes that lead to a CF are: (a) reduction in the three-phase a.c. bus voltages at the inverter; (b) increase in the d.c. current above its rated value; (c) harmonic presence in the a.c. voltages causing multiple voltage cross-overs; (d) pre-planned switching actions of filter banks and/or transformers and (e) lack of or insufficient firing pulse.

It is noted that the actual extinction angle measurement is available one cycle after the fault. In the case of equidistant firing control systems, only the mean value of 12 previous valve firings is used. Since a CF can occur at any instant, it is, therefore, not possible to predict a CF with 100% certainty. However, it may still be possible to predict the 'probability of a CF' under certain conditions, and in the knowledge of previously gained experience.

Fast controls are desired to reduce the impact of the CF on power flow. Traditional CF detection methods use (a) measurements of d.c. line voltage/current; (b) comparisons between the sum of the rectified three-phase a.c. currents into the converter and the d.c. line current; and (c) calculation of dv/dt and di/dt of the measured d.c. line voltage/current, respectively.

12.4.3 Firing Pulse Detectors

Converter thyristors are triggered with firing pulses that arrive in a sequential pulse train synchronised to the a.c. bus voltage with the aid of a phase-locked oscillator. In a steady-state operation, the pattern of the firing pulses is of a fixed duration and regular in nature; therefore, it is easy to classify. However, during fault conditions, the a.c. bus voltage may become distorted or even disappear altogether. Since the firing pulse train remains synchronised to the centre frequency of the free-running oscillator, the firing pulse pattern remains regular but shifts in phase.

However, owing to the requirements of the thyristor valve protection units when the valves are conducting fault currents, the duration of these firing pulses may also become elongated. A combination of the varying phase shift and pulse elongation results in an irregular pattern that is more difficult to classify. A trained NN offers many advantages for the indication of any error in the firing pulse unit, (i.e. false sequences, missing pulses, low strength of firing pulses, noise pollution, etc). Such FDD units can be extremely useful during commissioning as well as in operational modes.

12.4.4 Non-linearities in Components

The performance of HVDC systems in particular, and a.c. systems in general, are affected by various non-linear power system components such as transformers and surge arresters. Such components have characteristics that are known and predictable. NNs have been applied to fault detection in transformers. The presence of harmonics and inter-harmonics can provide clues to the mal-operation of certain equipment, or even guidance in the selection of operational configurations.

12.4.5 Converter Faults

Each converter valve is comprised of many tens of thyristors connected in series. Each thyristor is served by several passive components [10] to ensure voltage sharing and to protect individual thyristors from overvoltage, excessive dv/dt and di/dt stresses (Figure 12.5). The saturating reactors protect the thyristor from damage immediately after firing. Direct voltage is equalised across the thyristor string by a d.c. grading resistor that acts as a voltage divider. Dynamic voltages are equalised by a R-C snubber circuit. And a capacitive grading branch is used to protect against switching surges.

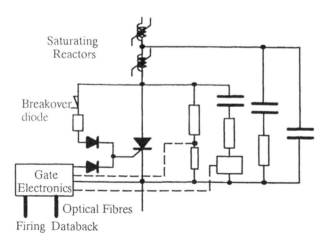

Figure 12.5 Electrical circuit of one thyristor level in converter valve design [10]

The command to fire a thyristor is emitted by the valve base electronics (VBE) unit at earth potential and fed via a fibre-optic cable to each thyristor operating at high voltage. The energy to fire the thyristor is derived from the grading circuit during the off-state interval.

Thyristors can be damaged easily by excessive forward voltage or forward dv/dt, especially at the elevated junction temperatures that occur during faults.

They are particularly vulnerable during the recovery period immediately after turn-off when even a modest forward voltage may cause uncontrolled conduction. Protection is afforded by firing the thyristor into conduction independently from the main control system. In marginal cases, some thyristors may block the forward voltage whilst others may not; in which case, the blocking thyristors would experience excessive voltages and suffer the consequences. A protective firing system, based on a breakover diode (BOD), is used to re-trigger the thyristor under such circumstances. The BOD can operate repeatedly in the case of failure of the VBE.

NNs can be used to classify the firing patterns emanating from the controllers, and monitor the voltages across each thyristor valve to indicate faults.

12.5 TEST CASES

Usually, difficulty is experienced in discriminating between a line-to-line fault (LLF) and double line-to-ground faults (DLGF). In Method 1, suggested in [1], the NN of Figure 12.6 was unable to discriminate between these two types of fault (Figure 12.7).

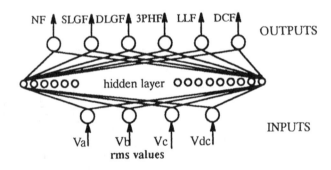

Figure 12.6 NN for test case of Method 1

However, in Method 2 [1] additional information, i.e. phase angles between the three phases is provided to enable the NN to make its decision. The concept of using the phase shift of ± 120° between the three phases is novel. Hence, one additional NN per phase (Figure 12.8) is needed to detect these faults. These phase detector networks are fed with the following inputs per phase (i.e. for phase A): (a) rms phase voltage, i.e. Va, in pu; (b) angle between phases A and B, i.e. Ph A-B, in pu; (c) angle between phases A and C, i.e. Ph A-C, in pu. At its output, one processing element is attached in an additional layer. This element is connected with fixed weights such that the output from the phase detector network is (PDN): (a) 1, when there is no fault (NF); (b) -1, when there is a line-to-line fault (LLF); (c) 0, when there is a single-line-to ground fault (SLGF). These outputs (X,Y and Z) for the three phases are then connected to the main NN. In Figure 12.9, the results of this NN test case are shown and this time the correct distinction is made.

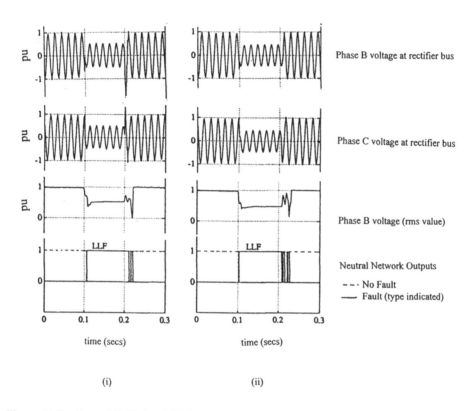

Phase B voltage at rectifier bus

Phase C voltage at rectifier bus

Phase B voltage (rms value)

Neutral Network Outputs

‒ ‒ ‐ No Fault
—— Fault (type indicated)

time (secs)   time (secs)

(i)   (ii)

Figure 12.7   Case of (i) LLF and (ii) DLGF with Method 1 (Figures 5 a & b from [1], reproduced by permission by the IEEE)

12.6 CONCLUSIONS

This chapter has focused on FDD applications in HVDC systems since such systems are fast-acting and can benefit from real-time fault monitoring, diagnosis, control and protection [6]. The chapter has been limited to FDD aspects only; the control and protection aspects have not been fully explored here.

Results from the NN-based monitoring techniques are very encouraging, as indicated by hardware implementations described elsewhere [11,12]. The recent implementation of a complete station-based FDD system [13] points to promising future industrial applications.

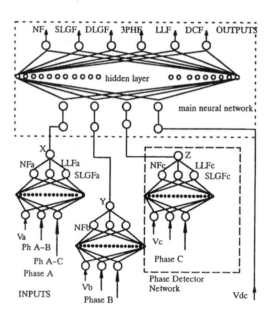

Figure 12.8   NN for test case of Method 2

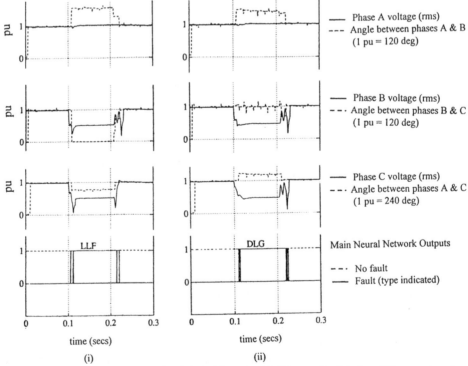

Figure 12.9   Case of (i) LLF and (ii) DLGF with Method 2 (Figures 7 a & b from [1], reproduced by permission by the IEEE)

Intelligent systems are still evolving in practice. A recent development is in the use of constructive Radial Basis Function (RBF) NNs. These yield the most optimum size of network, with the network development taking place during the training of the NN, commencing from the initial configuration having only two neurons in the hidden layer. The main requirement, from the FDD point of view, is that the NN configuration should be unambiguous, fast, able to operate even on corrupted data and able to incorporate system growth. Categorical identification of the type of fault and its location is also very important. More particularly, the FDD of evolving faults and its real-time operation will be an essential future requirement.

A particular challenge for the control and protection applications in an HVDC converter remains in the area of prediction and detection of a commutation failure (CF). The early prediction of a CF can be extremely useful in enhancing the dynamic performance of a HVDC system. The early prediction of a CF will require considerable understanding of the fundamental CF process in a thyristor valve. Such an understanding does not exist at present, and it is therefore an ideal application for a NN-based technique.

12.7 REFERENCES

[1] N Kandil, V K Sood, K Khorasani and R V Patel, 'Fault identification in an AC-DC transmission system using Neural Networks', *IEEE Transactions on Power Systems*, 7, 1992, 812-819.

[2] V K Sood, N Kandil, R V Patel and K Khorasani, 'Comparative evaluation of neural-network-based and PI current controllers for HVDC transmission', *IEEE Transactions on Power Electronics*, **9**, 1994, 288-296.

[3] L L Lai, F N Che, K Swarup and H Chandrasekharaiah, 'Fault diagnosis for HVDC systems using neural networks', *12th IFAC World Congress*, Australia, July 1993.

[4] L L Lai, F Ndeh-Che and T Chari, 'Fault identification in HVDC systems with neural networks', *IEE Second International Conference on Advances in Power System Control, Operation and Management*, 1993, 231-236.

[5] R Jayakrishna, H S Chandrasekharaiah and K G Narendra, 'Neuro-fuzzy controller for enhancing the performance of extinction angle control of inverters in a MTDC-AC system', *2nd International Forum on Applications of Neural Networks to Power Systems*, Y Tamaura, H Suzuki and H Mori, Editors, Yokohama, Japan, April 1993.

[6] V Shyam, H S Chandrasekharaiah and L L Lai, 'Real-time intelligent control for a multi-terminal direct current transmission system', *Proceedings of the Sixth European Conference on Power Electronics and Applications*, **1**, EPE, Sept 1995, 559-564.

[7] V K Sood, H S Chandrasekharaiah and L L Lai, 'Fault diagnosis using neural networks in HVDC systems', *Australian Journal of Intelligent Information Processing Systems*, **3**, Autumn 1996, 46-56.

[8] I Kamwa, R Grondin, V K Sood, C Gagnon, V T Nguyen and J Mereb, 'Recurrent neural networks for phasor detection and adaptive identificiation in power system control and protection', *IEEE Transactions on Instrumentation and Measurement*, **45**, 1996, 657-664.

[9] K Narendra, V K Sood, R Patel and K Khorasani, 'A neuro-fuzzy VDCL unit to enhance the performance of an HVDC system,' *IEEE Canadian Conference on Electrical and Computer Engineering*, Montreal, 1995.

[10]J D Wheeler and J L Haddock, 'Chandrapur back-to-Back HVDC scheme in India', *Proceedings of the International Conference on Power System Technology*, 1994, International Academic Publishers, 742-747.

[11]T Dalstein and B Kulicke, 'Neural network approach to fault classification for high speed protective relaying', *IEEE Transactions on Power Delivery*, **10**, 1995, 1002-1011.

[12]T S Sidhu, H Singh and M Sachdev, 'Design, implementation and testing of an artificial neural network nased fault direction discriminator for protecting transmission lines', *IEEE Transactions on Power Delivery*, **10**, 1995, 697-706.

[13]H Yang, W Chang and C Huang, 'On-line fault diagnosis of power substation using connectionist expert system', *IEEE Transactions on Power Systems*, **10**, 1995, 323-331.

An ANN Approach to the Diagnosis of Transformer Faults

This chapter presents an artificial neural network (ANN) approach to the diagnosis and detection of incipient faults in oil-filled power transformers on the basis of dissolved gas-in-oil analysis. An ANN method is used to detect faults with or without the involvement of cellulose. Accurate diagnosis is obtained with the proposed approach.

13.1 INTRODUCTION

Transformers are important pieces of equipment in electrical systems. Accurate fault diagnosis is vital for the reliable operation of these systems. Among the main diagnostic tools, dissolved gas analysis (DGA) of transformer oil has been the most useful technique to detect faults at an early stage. Transformers are subject to electrical and thermal stresses, which can break down the insulating materials and release gaseous decomposition products. Overheating, corona and arcing are three primary causes of fault-related gases. Principally, the fault-related gases commonly used are hydrogen (H_2), carbon monoxide (CO), carbon dioxide (CO_2), methane (CH_4), acetylene (C_2H_2), ethane (C_2H_6), and ethylene (C_2H_4). The analysis of dissolved gases is a powerful tool to diagnose developing faults in power transformers. Many diagnostic criteria have been developed for the interpretation of the dissolved gases. These methods find the relationship between the gases and the fault conditions, some of which are obvious and some of which may not be apparent. However, much of the diagnostics relies on experts to interpret the results correctly. New computer-aided techniques can consistently diagnose incipient-fault conditions and in some cases may provide further insight to the expert. Expert system and fuzzy set approaches have been developed to reveal some hidden relationships in transformer fault diagnosis [1-

5]. The expert system derives the decision rules from the previous experience while the fuzzy set represents the decision rules by using vague quantities. Although DGA has been used widely in the industry, in some cases the faults cannot be accurately determined. The IEC ratio codes, which provide fault diagnosis directly from a chromatography detector record, are the most widely used in the world. The IEC ratio codes are defined from the sharply definite gas ratio values and a ratio code will suddenly change in the neighbourhood of its boundaries. In practice, the gas ratio boundary may not be clear, especially when more than one type of fault exists. To overcome this problem, fuzzy set theories can be applied to the IEC ratio codes to produce fuzzy boundaries between different codes. The artificial neural network method (ANN) [5-6] has also been used to search for hidden relationships between the fault types and dissolved gases through training process. In this chapter, the use of fuzzy sets to diagnosis transformer faults will be presented first and then an ANN approach will be introduced. The accuracy of the ANN is carefully verified. With two ANNs, high diagnostic accuracy is obtained. In conclusion, this chapter aims to point towards the direction of integration of the two approaches, namely, fuzzy logic and ANNs.

13.2 DISSOLVED GAS ANALYSIS

Different patterns of gases are generated owing to the different intensities of energy dissipated by various faults. Totally or partially dissolved into the oil, the gases present in an oil sample make it possible to determine the nature of the fault by the gas types and their amount. The dissolved gas analysis (DGA) method is a widely used technique for transformer fault diagnosis. One of the DGA methods was derived from Halstead's discovery. Halstead made a theoretical thermodynamic assessment of the formation of the simple decomposition of gaseous hydrocarbons [1]. It states that the rate of evolution of any particular gaseous hydrocarbon varies with temperature, and that at a particular temperature there would be a maximum rate of evolution of that gas. Study of the Halstead thermodynamic equilibrium suggested that with increasing temperature, the hydrocarbon gas with a maximum rate of evolution would, in turn, be methane, ethane, ethylene and acetylene. Halstead's work proved the existence of a relationship between the fault temperature and the composition of dissolved gases. Efforts have been made to create simplified diagnostic criteria such as the key gas method and the ratio method, which, in essence, are based on this variation in gassing characteristics with the temperature to which the materials are subject.

13.2.1 Key Gas Method

Characteristic 'key gases' have been used to identify particular fault types [2]. The suggested relationship between key gases and fault types is summarised as:

H_2 corona
O_2 and N_2 non-fault related gases
CO and CO_2 cellulose insulation breakdown
CH_4 and C_2H_6 low temperature oil breakdown
C_2H_4 high temperature oil breakdown
C_2H_2 arcing

Excluding O_2 and N_2, there are seven fault-related gases. The fault condition is indicated by the excessive generation of these gases. Since this method does not give the numerical correlation, the diagnosis depends greatly on experience and therefore this technique is simple, yet labour intensive.

13.2.2 Ratio Methods

Rogers, Dornenberg and IEC are the most commonly used ratio methods. They employ the relationships between the gas contents. Key gas ppm values are used in these methods to generate the ratios between them. The ranges of the ratios are assigned to different codes that determine the fault types [7, 8]. Coding is based on experience and is always under modification. Ratio methods are limited in discerning problems when more than one type of fault occurs simultaneously.

The fuzzy set approach and expert systems have been used to incorporate various rules. A knowledge base or a fuzzy membership function is selected on the basis of past experience. The fault diagnosis is a weighted conclusion drawn from a number of data pertinent to the equipment. Its reliability increases with the amount of information available from previous tests and the degree of experience of the laboratory performing the analysis. Therefore, the knowledge base required could be large and complex.

13.3 IEC CODES FOR DISSOLVED GAS ANALYSIS

In dissolved gas analysis, IEC codes have been used for about two decades. Considerable experience has accumulated throughout the world to diagnose successfully incipient faults in the HV equipment with oil insulation. Early interpretations concentrated on specific gas components such as hydrogen and methane for the determination of discharges in oil. This simplistic approach was refined by a number of researchers [7, 8] who investigated the ratios of pairs of dissolved gases to set up a more extensive diagnostic technique. This technique was standardised by IEC in 1978 in the *Guide for Interpretation of the Analysis of Gases in Transformer and Other Oil Filled Electrical Equipment in Service* [9]. The individual gases used to determine each ratio and its assigned limits are shown in Tables 13.1 and 13.2. Codes are then allocated according to the value obtained for each ratio and the corresponding faults characterised.

Although the IEC code method is useful for the assessment of transformer insulation, no quantitative indication for the possibility of each fault is given. Also, in some cases, no suitable matching codes can be obtained, making the diagnosis unsuccessful. In multiple fault conditions, gases from different faults are mixed up, resulting in confusing ratios between different gas components. This can only be dealt with by more sophisticated analysis methods such as fuzzy diagnosis.

Table 13.1 IEC ratio codes

	IEC Codes		
Range of Gas Ratio		Code of Ratio	
	$\dfrac{C_2H_2}{C_2H_4}$	$\dfrac{CH_4}{H_2}$	$\dfrac{C_2H_4}{C_2H_6}$
<0.1	0	1	0
0.1-1	1	0	0
1-3	1	2	1
>3	2	2	2

Table 13.2 Fault characteristics from the IEC gas ratio codes

Fault No.	Fault Characteristics	$\dfrac{C_2H_2}{C_2H_4}$	$\dfrac{CH_4}{H_2}$	$\dfrac{C_2H_4}{C_2H_6}$	Typical Examples
0	No fault	0	0	0	Normal ageing
1	Partial discharges of low energy density	0 (but not significant)	1	0	Discharges in gas-filled cavities resulting from incomplete impregnation or super saturation or cavitation or high humidity
2	Partial discharges of high energy density	1	1	0	As above, but leading to tracking or perforation of solid insulation
3	Discharges of low energy	1-2	0	1-2	Continuous sparking in oil between bad connections of different potential or to floating potential. Breakdown of oil between solid materials
4	Discharges of high energy	1	0	2	Discharges with power follow through. Arcing-breakdown of oil between coils to earth. Selector breaking current
5	Thermal fault of low temperature < 150°C	0	0	1	Insulated conductor overheating

Table 13.2 continued

6	Thermal fault of low temperature 150 - 300°C	0	2	0	Local overheating of the core due to concentration of flux. Increasing hot spot temperatures, varying from small hot spots in core, shorting links in core, overheating of copper due to eddy currents, bad contacts/joints (pyrolytic carbon formation) up to core and tank circulating currents
7	Thermal fault of medium temperature 300 - 700°C	0	2	1	
8	Thermal fault of high temperature > 700°C	0	2	1	

13.4 FUZZY SETS OF IEC GAS RATIO CODES

According to the IEC codes in Table 13.1, the three gas ratios C_2H_2/C_2H_4, CH_4/H_2 and C_2H_4/C_2H_6 can be coded as 0, 1 and 2 for different ranges of ratios. Table 13.1 is rearranged for a clear relationship between the range of each gas ratio and the codes, as shown in Table 13.3. This relationship can also be represented graphically; for example, for gas ratio $r = C_2H_4/C_2H_6$, its ratio codes are depicted in Figure 13.1(a).

Table 13.3 The rearranged IEC codes

Ratio Code	Code 0	Code 1	Code 2
$\dfrac{C_2H_2}{C_2H_4}$	<0.1	0.1-3	>3
$\dfrac{CH_4}{H_2}$	0.1-1	<0.1	>1
$\dfrac{C_2H_4}{C_2H_6}$	<1	1-3	>3

These three sets of IEC codes, 0, 1 and 2, can be reconstructed as fuzzy sets ZERO, ONE, TWO using fuzzy set theories. Any ratio value of $r = C_2H_4/C_2H_6$ can be represented as the following fuzzy vector:

$$[\mu_{\text{ZERO}}(r), \mu_{\text{ONE}}(r), \mu_{\text{TWO}}(r)]$$

where $\mu_{\text{ZERO}}(r), \mu_{\text{ONE}}(r), \mu_{\text{TWO}}(r)$ are the membership functions of fuzzy sets ZERO, ONE and TWO, respectively.

Figure 13.1(b) shows the membership functions $\mu_{\text{ZERO}}(r), \mu_{\text{ONE}}(r), \mu_{\text{TWO}}(r)$ for the gas ratio C_2H_4/C_2H_6.

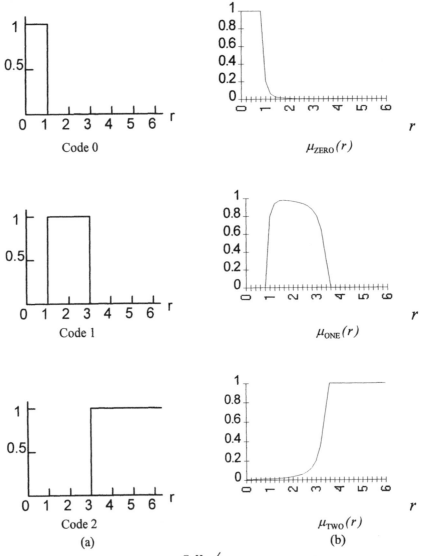

Figure 13.1 IEC codes for the gas ratio $r = C_2H_4/C_2H_6$ (a) and their fuzzy sets (b)

According to Table 13.2, transformer faults can be identified by the IEC codes of three gas ratios: $r_1 = {}^{C_2H_2}\!\big/\!{}_{C_2H_4}$, $r_2 = {}^{CH_4}\!\big/\!{}_{H_2}$ and $r_3 = {}^{C_2H_4}\!\big/\!{}_{C_2H_6}$. For example, if the code of ${}^{C_2H_2}\!\big/\!{}_{C_2H_4}$ is zero, the code of ${}^{CH_4}\!\big/\!{}_{H_2}$ is two and the code of ${}^{C_2H_4}\!\big/\!{}_{C_2H_6}$ is one, the fault of the transformer is predicted as the seventh fault; i.e. a thermal fault of medium temperature 300 - 700°C. In this fuzzy diagnosis method, the conventional IEC codes are replaced by the fuzzy sets of gas ratio codes, the logic 'AND' by the minimisation operation and the logic 'OR' by the maximisation operation. In other words, fuzzy multivalue logic is used in this method to substitute for conventional true-false logic [10].

13.5 FUZZY SETS OF KEY GASES

Characteristic 'key gases' have been used to identify particular faults. The faults are indicated by the excessive generation of the relevant gases. For different voltage levels, winding structures and applications of transformers, the threshold level may be different because the gas generated and dissolved in oil is dependent on many factors. According to fuzzy set theories and the experience acumulated using the conventional key gas method, the membership functions of fuzzy sets 'LOW', 'MEDIUM' and 'HIGH' can be represented as the descending demi-Cauchy distribution function $\mu_L(x)$ or the ascending demi-Cauchy distribution function $\mu_H(x)$ or their combination.

For every given key gas concentration x, the corresponding fuzzy vector is $[\mu_{LOW}(x), \mu_{MED}(x), \mu_{HIGH}(x)]$. The fuzzy IEC code-key gas method produces a fuzzy component which is a combination of fuzzy diagnosis using the IEC code and key gases [2].

13.6 RESULTS

In the conventional IEC method, deterioration of specific faults could be determined if only a single fault existed. With multiple faults in a transformer, dissolved gases from different faults are mixed up, making it difficult to analyse separately individual faults. In fuzzy diagnosis, a fault can be more accurately determined by its fuzzy component. Deterioration of this fault may therefore be closely monitored from the trend of its fuzzy component. For example, tests on the oil of a transformer were carried out 11 times during a 15-month period. Using the fuzzy IEC code-key gas method, a thermal fault (300-700°C and >700°C) was diagnosed as the main fault inside the transformer. The results are plotted in Figure 13.2 and show the trend of the thermal fault. It can be seen that the medium temperature thermal fault F(7) was the main cause of the insulation deterioration of the transformer. The possibility of high temperature thermal fault (>700°C) was very small. The high temperature thermal fault F(8) was diagnosed on about Day 114 and then become stable until Day 406 when the oil was de-gassed. Because the thermal fault remained after de-gassing, the fuzzy components F(7) and F(8) went up again from Day 453. A small fluctuation of F(8) was recorded on Day 178 which might have been due to the lighter load during the specific time period.

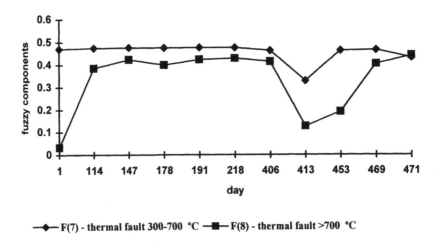

Figure 13.2 Trend of a thermal fault in a transformer determined by the fuzzy IEC code-key gas method

13.7 THE ARTIFICIAL NEURAL NETWORK

Very complex systems can be characterised with very little explicit knowledge using ANNs. In contrast, expert and fuzzy set systems can only use explicit knowledge to establish knowledge base and fuzzy membership function selection. Theoretically, a neural network could represent any observable phenomenon. The relationship between gas composition and incipient-fault conditions is learned by the ANN from actual experience (through training samples). Obvious and not so obvious (hidden) relationships are detected by the ANN and used to develop its basis for the interpretation of dissolved gas-in-oil data.

An ANN design includes the selection of input, output, network topology (structure, or arrangement of nodes) and weighted connections of the nodes. Input feature (information) selection constitutes an essential first step. The feature space needs to be chosen very carefully to ensure that the input features will correctly reflect the characteristics of the problem.

An ANN design can be done experimentally through a repeated process to optimise the number of hidden layers according to training performance and prediction accuracy.

13.8   THE ANN APPROACH TO FAULT DIAGNOSIS

Since there are already various diagnostic criteria in use, the significance of using ANN is to achieve a better diagnosis performance. It is also useful to distinguish fault in a cellulose from that in the oil.

Overheating, corona and arcing are three major fault types. Since each of them could involve cellulose breakdown, there could be seven patterns to be identified, including the normal condition.

Often, a large number of data is not available for ANN training and performance verification. Thus, cross-validation is used to train and test the ANN.

13.8.1   The ANN for Major Fault Type Diagnosis

Theoretically, the above diagnosis and detection problem can be realised by one ANN using all seven fault-related gases as inputs. It was found in a previous study that the ANN with CO_2 as one of the input feature had a worse performance in major fault type (arcing, corona, overheating) diagnosis than that without CO_2 [6]. CO and CO_2 are needed as the input features since they are the primary gases formed from the degradation of cellulose. For this reason, two ANNs are used to separate the cellulose condition detection from the major fault type diagnosis. One ANN is trained only to classify the major fault types while another focuses on determining whether the cellulose is involved.

To examine major fault type diagnosis, five key gases, H_2, H_4, C_2H_2, C_2H_6 and C_2H_4, are chosen as input features. Both the cases with and without 'CO' as input features are studied. The cases with and without 'normal' output node are also considered. In total, there are four combinations to be tested and compared. The ANN is trained using the conventional backpropagation method. Since the ANN topology has a great impact on classification accuracy and convergence time during training, several network topologies are compared. Usually, a one-hidden-layer ANN is enough for most non-linear mapping. However, ANNs with one, two and 3 hidden layers are trained and compared. According to the number of training iterations and training errors, the optimal is determined. The cross-validation of this optimal ANN is computed to verify the accuracy. Forty sample sets from different transformers are used for training and testing the ANNs.

Comparing the accuracy, iteration number and network complexity, it is shown that the two-hidden-layer ANN topology is the optimal for all four cases. An input feature space without CO increases the accuracy. Keeping 'normal' as an output node also increases the accuracy. The optimal ANN has a 95% chance of success in classifying the three major fault types. The outputs for the ANN are normal, overheating, corona, and arcing.

13.8.2 The ANN for Cellulose Condition Detection

Another ANN is constructed with one output to test only the cellulose condition. Output '1' means that cellulose is involved in the faults, while '0' represents no cellulose involved. Although only CO and CO_2 contain the information about the cellulose condition, their ratios to the hydrocarbon gases provide useful information in detecting the cellulose condition. Together with CO and CO_2, five hydrocarbon key gases are also used as input features. According to [6], the two-hidden-layer ANN is the optimal one that gives about 90.9% accuracy.

Since the CO_2 to CO ratio is an indicator of the cellulose condition, this ratio is used directly as one of the inputs in addition to the actual ppm values of CO and CO_2 in the ANN training. It is found that the rate of convergence does improve, while the overall level of diagnostic accuracy remains the same for the samples tested.

13.9 ANN DIAGNOSIS RESULTS

Two ANNs were constructed according to the above evaluation. The weight matrices and biases were stored as files. Testing data were obtained from a data file. The two ANNs were used to detect the faults and the cellulose condition. Five new sample sets from different transformers were tested. The testing data and results are listed in Tables 13.4 and 13.5, respectively [6].

Table 13.4 Testing data (ppm) (Table 6 from [6], reproduced by permission by the IEEE)

Case	H_2	CH_4	C_2H_2	C_2H_4	C_2H_6	CO	CO_2
1	130	98	65	56	7	110	4000
2	300	240	140	160	14	23	160
3	17000	110000	16000	89000	84000	320	430
4	320	1370	9	1980	417	19	549
5	1400	3000	4	3500	560	1600	6800

The results between the actual inspection of the transformer and the ANN diagnosis match very well. There are two cases (1 and 2) where the ANN for the major fault type diagnosis and the actual inspection disagree. In these two cases, there is arcing but the ANN indicates that there is arcing (A) and overheating (O). The ANN can be trained with refined input data to distinguish between these two because the arcing degradation products are the same as the thermal degradation products but in a different relative composition. There are cases where there is overheating that causes damage to the insulation, which then results in electrical discharges.

Table 13.5 Testing results (0-1 range represents the different degree and certainty of a particular diagnosis) (Table 7 from [6], reproduced by permission by the IEEE)

Case	Inspection	Normal	Overheating	Corona	Arcing	Cellulose	ANN Diagnosis
1	Arcing (p)*	0.0000	0.9269	0.0003	0.9981	0.9342	A and O (p)*
2	Arcing	0.0000	0.8771	0.0001	0.9998	0.2123	A and O
3	Overheating (p)*	0.0000	1.0000	0.0000	0.0000	0.9997	Overheating (p)*
4	Overheating (p)*	0.0000	0.9999	0.0000	0.0000	0.5621	Overheating (p)*
5	Overheating (p)*	0.0000	0.9999	0.0000	0.0000	0.9768	Overheating (p)*

(p)* Paper or other cellulosic materials involved.

In Case 3, there is a large quantity of C_2H_2 (16000 ppm). A quick evaluation may suggest that there is arcing in the oil as many people are sensitive to the appearance of C_2H_2. However, the relative composition of the gases indicates that the problem is thermal in nature in agreement with the ANN output and as confirmed by the internal investigation. In this case, the amounts of carbon oxides is not high; however, the ratio of CO_2 to CO is quite high, indicating the possibility of the involvement of cellulosic materials in the arc as confirmed by inspection of the transformer.

For Case 4, it is difficult to justify the ANN evaluation of the overheating of paper even though some was found on inspection of the transformer. It would appear that the ANN is picking up on the low CO_2 to CO ratio, which is not always a good indicator of the overheating of cellulosic materials. It may be that some minimum quantity limits for these gases may be required. This can be implemented easily in the ANN computer code.

A number of cases for which the diagnosis indicates the involvement of cellulosic materials requires careful review. In Case 1, there is an indication of the overheating of paper both during the inspection of the transformer and by the ANN. There is an appreciable amount of CO_2 present. However, this is not an unusual amount and a significant portion is likely from the general ageing of the transformer cellulosic insulation rather than the fault. This is one of the difficulties when there is no previous data to show which gases are most actively being generated. This is similar in Case 5, where it is difficult to determine without any specific previous data whether the carbon oxides generated are from the incipient-fault location or from the normal ageing of the cellulosic materials. In Case 5, the investigation revealed loose connections resulting in hot metal and therefore it seems likely that the majority of the carbon oxides was not from the incipient-fault areas but rather from the ageing of the winding insulation.

13.10 DISCUSSION

The preliminary results have proved that by using the fuzzy diagnosis method for dissolved gas analysis, more detailed information about the multiple incipient faults inside a transformer can be obtained. This is a significant improvement over the conventional IEC code method. This is because the fault type is related in a more realistic way to the fuzzy key gas sets by the key gas concentration and to the gas ratio by the fuzzy IEC code set.

Theoretically, the ANN can be trained to represent any observable phenomenon if sufficient data are available. The more complex a relationship, the more training data are needed. Transformer fault diagnosis can be very complicated. For example, it is desirable to distinguish between faults of oil and cellulosic materials (paper), between different temperatures (low, medium and high) for overheating in oil, or between low energy and high energy sustained arcing. To deal with such a complicated diagnosis problem, the available input data may not always be enough. It can be a very effective way to construct different ANNs for different pattern recognition to obtain the highest degree of diagnostic accuracy for each pattern.

13.11 CONCLUSIONS

An ANN approach is promising for transformer fault diagnosis even with limited sample data. Patterns with different identification sensitivity are detected by different ANNs. Future investigations will involve the application of NNs to fine-tune the fuzzy model. On the basis of the test results from a certain type of transformer, the parameters of the fuzzy model can be more accurately determined which may significantly improve the accuracy and sensitivity of the diagnosis of multiple faults in a power transformer.

13.12 REFERENCES

[1] C E Lin, J. M. Ling and C. L. Huang, 'An expert system for transformer fault diagnosis and maintenance using dissolved gas analysis', *IEEE Transaction on Power Deliver*, **8**, 1993, 231-238.

[2] Q Su and C. Mi, 'Fuzzy diagnosis of transformers in dissolved gas analysis', *Research Report*, Centre for Electrical Power Engineering, Monash University, Australia, June 1996.

[3] K Tomsovic, M Tapper and T Ingvarsson, 'A fuzzy information approach to interpreting different transformer diagnostic methods', *IEEE Transactions on Power Delivery*, 8, 1993, 1638-1646.

[4] Y C Huang, H T Yang and C L Huang, 'Developing a new transformer diagnosis system through evolutionary fuzzy logic', *IEEE Transactions on Power Delivery*, 12, 1997, 761-767.

[5] Z Wang, Y Liu and P J Griffin, 'A combined ANN and expert system tool for transformer fault diagnosis', *IEEE/PES Winter Meeting, PE-411-PWRD-0-12-1997*, Feb 1998.

[6] Y Zhang, X Ding, Y Lin and P J Griffin, 'An artificial neural network approach to transformer fault diagnosis', *IEEE Transactions on Power Delivery*, 11, 1996, 1836-1841.

[7] J Kelly, 'Transformer fault diagnosis by dissolved-gas analysis', *IEEE Transactions on Industry Applications*, **IA-16**, 1980, 777-782.

[8] R Rogers, 'IEEE and IEC codes to interpret incipient faults in transformer, using gas in oil analysis', *IEEE Transactions Electrical Insul*ation, **EI-13**, 1978, 349-354.

[9] IEC Publication 599, 'Interpretation of the analysis of gases in transformers and other oil-filled electrical equipment in-service', 1978.

[10] L A Zadeh, 'Fuzzy sets', *Information and Control*, Academic Press, **8**, New York, 1965, 338-353.

14

Real-Time Frequency and Harmonic Evaluation Using Artificial Neural Networks

With increasing harmonic pollution in the power system, real-time monitoring and analysis of harmonic variations have become important. Because of the limitations associated with conventional algorithms, particularly under supply-frequency drift and transient situations, a new approach based on non-linear least squares parameter estimation has been proposed as an alternative solution for high-accuracy evaluation. However, the computational demand of the algorithm is very high and it is more appropriate to use Hopfield-type feedback neural networks (NNs) for real-time harmonic evaluation. The proposed NN implementation determines simultaneously the supply-frequency variation, the fundamental-amplitude/phase variation, as well as the harmonic-amplitude/phase variation. The distinctive feature is that the supply-frequency variation is handled separately from the amplitude/phase variations, thus ensuring a high computational speed and a high convergence rate. Examples by computer simulation are used to demonstrate the effectiveness of the implementation. A set of data taken on site were used as a real application of the system.

14.1 INTRODUCTION

The electrical power system has been increasing in complexity at a rapid rate in the last few decades. Many measures have been introduced to improve its reliability and security. However, the system is becoming more and more polluted owing to the increasing use of power-electronic converters and controllers for industrial processes and drives, among other types of disturbing loads [1]. The effect is the contamination of the 50 Hz supply by a wide range of frequencies up to the radio frequencies, and thus power system monitoring becomes a necessity to check the state of health of the power network [2]. Well-proven techniques have been devised to monitor various significant quantitative measures, such as r.m.s voltage, r.m.s current, the power factor and active and reactive power, under both steady and transient states. However, there has been difficulty in tracking frequency if the signals are not clean. Harmonic monitoring is still considered not well developed [3]. Discrete Fourier Transform (DFT) has been a well-known method for frequency spectra evaluation [4]. The main reason for the success of the DFT-based techniques for spectral analysis is the low computational requirement. The use of Fast Fourier Transform (FFT) algorithms reduces the computational time required for the evaluation of the DFT by several orders of magnitude. The method achieves this efficiency by eliminating redundancies appearing in the evaluation of different DFT coefficients. However, there are several inherent performance limitations in the DFT approach and they are due to the implicit windowing of the data that occurs when processing blocks of data with FFT. Windowing manifests itself as 'leakage' in the spectral domain [5]. Spectral leakage occurs when the time record data used by the FFT algorithm does not contain an exact integral number of power-frequency cycles or if it contains frequency components that do not correspond to one of the spectral lines. Since the power system frequency is subject to small random deviations, some degree of spectral leakage inevitably occurs. Moreover, DFT techniques for spectral analysis are inherently limited in terms of resolution (or in the extent to which individual frequency components may be resolved) [5] and the fundamental frequency needs to be pre-assumed. These two performance limitations of the DFT approach are particularly evident when analysing short data sequences.

Recent studies have reported improvements in DFT-based procedures that first estimated the frequency and then applied the DFT using a window that was a multiple of the estimated period. A variety of numerical algorithms dedicated to frequency measurement have been published in the last decade. Zero-crossing detection and calculation of the number of cycles that occur in a predetermined time interval is a simple and well-known method [6,7]. Although the use of zero crossing is simple and quite reliable even under low-level noise contamination [7], the level of accuracy will be lowered in cases of very high-level harmonic distortion and the estimation time period is quite long since a multiple number of full cycles must be available before any estimation can be carried out. One method made use of the Taylor series technique and digitised samples of voltage at a relaying point [8]. Such a method suffers the same problems as the zero-crossing method. Another method made use of DFT as the starting point to estimate the fundamental phase angle recursively and then the fundamental frequency was fine-tuned on the basis of changes in the shifting of phase angles [9,10]. This method will only be successful if the fundamental frequency drift is very small. Otherwise, the spectral leakage has already imposed a very large error on the initial estimation of phase angles. The methods of linear least squares [11] and Kalman filtering [12] were developed. In most cases, the methods only considered the fundamental component and hence the existence of higher harmonic would seriously affect the results. The latest approach made use of non-linear curve fitting to estimate both the fundamental frequency and the higher harmonic [13,14]. Applying iterative Newton's procedure combined with an ordinary least-squares technique, high measurement accuracy over a wide range of frequency changes and very fast algorithm convergency can be achieved. However, it is very computationally intensive and demanding and thus, parallel computers must be employed for real-time calculation. This is the major reason why the ANN is called in.

The implementation of the NN in harmonic monitoring is not a new concept [15,16]. This chapter presents the use of the least-squares technique to harmonic extraction in time-varying situations with significant advantages together with the implementation of ANNs. This system operates in real time with a very fast response and can handle situations where the information is not complete.

14.2 NON-LINEAR LEAST SQUARES

The application of the least-squares approach to time-varying frequency and harmonic extraction is examined below. The least-squares approach is basically fitting the sampled waveform to a harmonic equation under the minimum total squared error criterion. Consider the following Fourier expansion of a current or voltage wave in the power system with n harmonic terms:

$$
\begin{aligned}
y(t) &= \sum_{k=0}^{n} A_k \sin\left(2\pi kft + \theta_k\right) \\
&= \sum_{k=0}^{n} \left(a_k \sin 2\pi kft + b_k \cos 2\pi kft\right)
\end{aligned}
\tag{1}
$$

Assume that there are m measuring samples, (t_i, y_i), where $i = 0$ to $m-1$. The objective is to estimate the values of a_k, b_k and f in (1) so as to minimise the total squared error, E:

$$
E = \sum_{i=0}^{m-1} \left(y_i - \sum_{k=0}^{n} \left(a_k \sin 2\pi kft_i + b_k \cos 2\pi kft_i\right) \right)^2
\tag{2}
$$

By applying Taylor's series expansion on E, Newton's method [17] can be used to solve the optimisation problem. Let $x=[b_0 \ a_1 \ b_1 \ ... \ a_n \ b_n \ f]^T$ be the vector of variables of the system being monitored. The iterative algorithm is then given as follows:

$$
x_{r+1} = x_r - H(x_r)^{-1} g(x_r)
\tag{3}
$$

Newton's method offers high rates of convergence. If the functions are quadratic, it is possible to arrive at the minimum in a single step. However, the excessive computational requirement for obtaining the inverse of the Hessian matrix $H(x)$ makes the algorithm rather unsuitable for real-time applications.

Therefore, the parallel operational capability of NNs has to be used to speed up the applications.

14.3 NN FORMULATION

As mentioned in the previous section, the computation of the Hessian matrix is very intensive, so the dynamic gradient approach is employed with the aid of NNs because it is not required to estimate **H(x)**.

14.3.1 The Dynamic Gradient System

It is possible to find the minimum of E by optimising on all variables but that will also be time-consuming and global convergence cannot be guaranteed. In this way, **x** is segmentated into two parts, i.e. the amplitude vector, **m**, and the frequency scalar, f; $\mathbf{x} = [\ \mathbf{m}^T\ ,\ f\ ]^T$. Therefore, the energy function, $E(\mathbf{x})$, can be minimised with respect to the vector **m** and scalar f along the optimisation time scale, t, by implementing a dynamic gradient system [18], as shown:

$$\frac{d\mathbf{m}}{dt} = -\mu_m \nabla E\,(\mathbf{m})\ ;\ \ \frac{df}{dt} = -\mu_f \nabla E(f) \tag{4}$$

where μ_m = diagonal step - size matrix

$\quad\ \ \mu_f$ = frequency step size

For illustrative purposes, three typical elements of ∇E are derived as shown:

$$\frac{\partial E}{\partial a_k} = 2\,\mu_{a_k} \sum_{i=0}^{m-1}\left(\ y_i\ -\ \Sigma(\cdot)\ \right) \sin 2\pi k f\, t_i$$

$$\frac{\partial E}{\partial b_k} = 2\,\mu_{b_k} \sum_{i=0}^{m-1}\left(\ y_i\ -\ \Sigma(\cdot)\ \right) \cos 2\pi k f\, t_i \tag{5}$$

$$\frac{\partial E}{\partial f} = 4\pi\,\mu_f \sum_{i=0}^{m-1}\left(\ y_i\ -\ \Sigma(\cdot)\ \right) R$$

where

$$\Sigma(\cdot) = \sum_{k=0}^{n}\left(a_k \sin 2\pi k f\, t_i + b_k \cos 2\pi k f\, t_i\right)$$

$$R = t_i \sum_{k=0}^{n} k\left[-b_k \sin 2\pi k f\, t_i + a_k \cos 2\pi k f\, t_i\right]$$

The constants, 2 and 4π, can be absorbed into the corresponding step-size parameters.

14.3.2 Neural Network Implementation

From (4), it can be seen that optimisation comes with two major blocks, the first block handling **m**, while the second block handles f. The first block is actually a linear optimisation by assuming that f is given. Initially, f is set to the nominal line frequency, i.e. 50 Hz. With the given f, **m** is evaluated by the first block and the estimated **m** is fed to the second block for evaluating f. t_i is continuously incrementing with the measured samples, y_i, and the two NN-based blocks are exchanging **m** and f and the process is repeating until the optimised **m** and f are obtained. The self-explanatory block diagram of the whole NN is shown in Figure 14.1. The z^{-1} module is a time delay operator for handling numerical integration; the $\otimes$ module serves as a multiplier; the $\oplus$ module serves as an adder, while the two rectangular blocks are identical, preparing the updated sine and cosine tables. All a s, in the neural network, although different from one another, are kept constant throughout the whole learning and optimisation process.

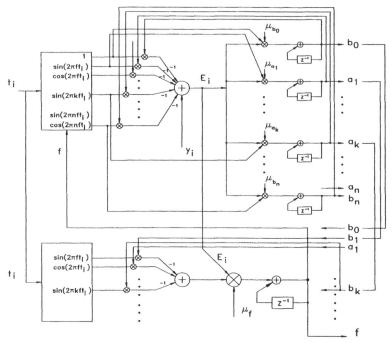

Figure 14.1 The NN system for frequency and harmonic evaluation

The general NN consists of two subneural networks. The first subneural network has $(2n+1)$ output nodes and three input nodes. The second subneural network has one output node and $(2n+3)$ input nodes. Initial values are application-specific, e.g. $b_0 = 1$, $a_1 = 220\sqrt{2}$ and other a s, b s $= 0$ for cases of voltage evaluation.

14.4 MODEL VALIDATION BY SIMULATION

To show the capability of the proposed system, a frequency-modulated (with f fluctuating sinusoidally at 2 Hz between 46 Hz and 54 Hz), fundamental amplitude-varying (amplitude exponential decaying) and harmonic amplitude-changing (step change in the fifteenth harmonic amplitude at 0.25 s from one unit to two units) waveform is arbitrarily generated. The instantaneous waveform with a 4.8 kHz sampling rate is shown in Figure 14.2 and the results of monitoring are shown in Figures 14.3 to 14.7. The specifications are given below.

f fluctuating sinusoidally at 2 Hz between 46 Hz and 54 Hz
A_1 decreasing from 300 V peak according to e^{-2t}
A_{15} abruptly changing from 1 to 2 V peak values
30 ms measurement window
$k = 1, 3, ..., n = 21$

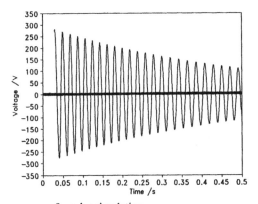

Figure 14.2   Instantaneous waveform by simulation

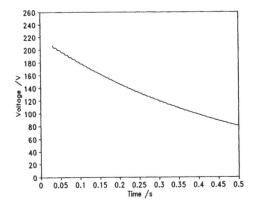

Figure 14.3   Extracted fundamental amplitude

The trends are accurately reproduced provided that the fifteenth harmonic component is taken into account. It can be seen that the NN system can accurately and efficiently trace the variations in fundamental frequency, which can never by done by conventional DFT, as well as the variations in amplitude of each harmonic content. If DFT only is employed to detect the fundamental and fifteenth harmonic amplitude, Figures 14.6 and 14.7 will be produced, indicating a tremendous error in identifying the fundamental and fifteenth harmonic accurately.

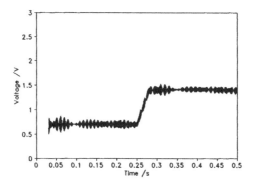

Figure 14.4   Extracted fifteenth harmonic amplitude

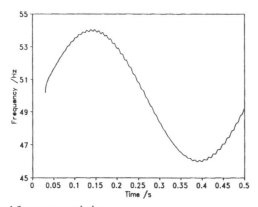

Figure 14.5   Extracted frequency variation

14.5   REAL APPLICATION OF THE SYSTEM

To show that the system can keep track of the real-world data to perform frequency and harmonic evaluations, the system is installed on site at a commercial centre in Hong Kong and a real-time waveform of 0.5 s is shown in Figure 14.8. The major load is a VVVF drive for an air-handling unit, together with other power electronic loads such as personal computers and electronic gears for fluorescent lamps. Again, a 30 ms measurement window is adopted for the entire process. The length of the measurement window is arbitrary under our algorithms and this is one of the merits of the whole system. From Figure 14.9, it can be seen that there is a slight variation of r.m.s. phase voltage over the period of monitoring but, in general, the voltage can be kept constant at around 223 V. Figure 14.10 shows the amplitudes of the third and fifth harmonic by the NN system while Figure 14.11 shows the variation in main frequency. It should be noted that there is continuous variation in the main frequency and the absolute value is not at 50 Hz (49.92 Hz, on average), which cannot be handled by the conventional DFT approach. Figures 14.12 and 14.13 show the harmonic spectrum in terms of both amplitude and phase at time 0.5 s.

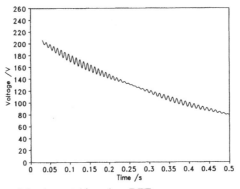

Figure 14.6 Extracted fundamental based on DFT.

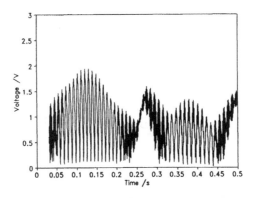

Figure 14.7 Extracted fifteenth harmonic amplitude based on DFT

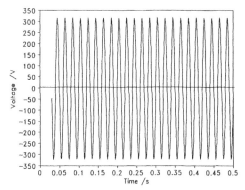

Figure 14.8 Instantaneous supply voltage waveform

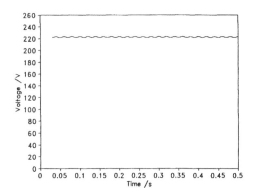

Figure 14.9 Extracted fundamental voltage by NN

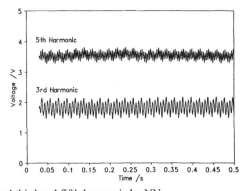

Figure 14.10 Extracted third and fifth harmonic by NN

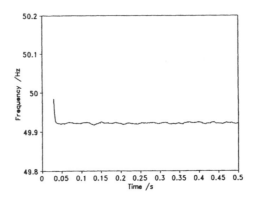

Figure 14.11   Extracted main frequency by NN

14.6 CONCLUSIONS

Conventional FFT techniques based on a fixed measurement window are not suitable for effective power system monitoring where there are frequency drifts in the signals involved. The proposed method is designed to deal simultaneously with the measurement of the varying frequency, main-component amplitude and any harmonic component(s) present in the continuously changing power system conditions. There is no theoretical restriction in the number of harmonic components to be evaluated except that the complexity of the neural network will be increased as the number of harmonic components is increased. The computational loading simply increases rapidly as the expected number of harmonic becomes large. This imposes constraint on real-time tracking and can be solved either by parallel processing or by the solution becoming obvious as the computational speed improves in the near future because of advances in technology. In this chapter, state-of-the-art technology in the ANN has been employed. The estimation of all a s and b s and f in the network is parallel and thus a high computational speed can be attained. The separation of f from a s and b s into two NNs is a very particular choice because this will highly simplify the equation set, reducing the high non-linearity of the whole system. This action not only simplifies the design of the network but also greatly improves the rate and probability of convergence.

The two examples presented here, one by simulation and one by on-site real measurement, illustrate the reliability of the method in extracting the multiaspect information in one single operation. The major varying features of the signals are truly deduced by the proposed method. The results are no longer critically dependent on the choice of the measurement window. In many power applications, only the mains frequency and the fundamental voltage amplitude are of concern in the monitoring. Provided that harmonic pollution is the only form of contamination in the system, and the highest significant order of harmonic present can be established with confidence beforehand, the frequency and voltage information can be extracted by the proposed system with high fidelity on a real-time basis.

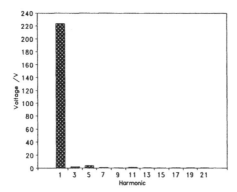

Figure 14.12 Extracted voltage harmonic spectrum at *t*=0.5 s

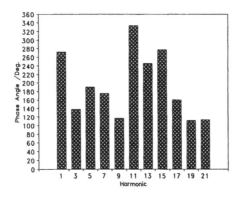

Figure 14.13 Extracted phase spectrum at *t*=0.5 s

14.7 ACKNOWLEGMENTS

The author wishes to acknowledge the IEEE for granting permission to reproduce the material and results contained in [19].

14.8   REFERENCES

[1]   T H Ortmeyer, K R Chakravarthi and A A Mahoud, 'The effects of power system harmonics on power system equipment and loads', *IEEE Transactions on Power Apparatus and Systems*, **104**, 1985, 2555-2563.

[2]   T A George and D Bones, 'Harmonic power flow determination using the fast fourier transform', *IEEE Transactions on Power Delivery*, **6**, 1991, 530-535.

[3]   J Arrillaga, D A Bradley and P S Bodger, *Power System Harmonics*, John Wiley & Sons, 1985.

[4]   J T Broch, *Principles of Experimental Frequency Analysis*, Elsevier Science Publishers Ltd., 1990, ch 4.

[5]   A A Girgis and F M Ham, 'A qualitative study of pitfalls in FFT', *IEEE Transactions on Aerospace and Electronic Systems*, **AES 16**, 1980, 434-439.

[6]   S A McIlwaine, C E Tindall and W McClay, 'Frequency tracking for power system control', *IEE Proceedings Pt. C*, **133**, 1986, 95-98.

[7]   M M Begovic, P M Djuric, S Dunlap and A G Phadke, 'Frequency tracking in power networks in the presence of harmonics', *IEEE Transactions on Power Delivery*, **8**, 1993, 480-486.

[8]   L L Lai, A D Wang, Y Z Ge, A O Ekwue and K L Lo, 'A simple method for measuring power system frequency', *Proceedings of the Third International Symposium on Electricity Distribution and Energy Management*, **2**, IEEE Singapore Section, 1993, 784-789.

[9]   A G Phadke, J S Thorp and M G Adamiak, 'A new measurement technique for tracking voltage phasors, local system frequency and rate of change of frequency', *IEEE Transactions on Power Apparatus and Systems*, **102**, 1983, 1025-1038.

[10] Working Group H-7 of the Relaying Channels, 'Synchronized sampling and phasor measurements for relaying and control', *IEEE Tranactions on Power Delivery*, **9**, 1994, 442-452.

[11] M S Sachdev and M M Giray, 'A least error squares technique for determining power system frequency', *IEEE Transactions on Power Apparatus and Systems*, **104**, 1985, 437-443.

[12] H M Beides and G T Heydt, 'Dynamic state estimation of power system harmonics using kalman filter methodology', *IEEE Transactions on Power Delivery*, **6**, 1991, 1663-1670.

[13] S K Tso and W L Chan, 'Frequency and harmonic evaluation using non-linear least-squares technique', *Journal of Electrical and Electronics Engineering*, **14**, 1994, 124-132.

[14] V Terzija, M Djuric and B Kovacevic, 'A new self-tuning algorithm for the frequency estimation of distorted signals', *IEEE Transactions Power Delivery*, **10**, 1995, 1779-1785.

[15] R K Hartana and G H Richards, 'Harmonic source monitoring and identification using neural networks', *IEEE Transactions on Power Systems*, **5**, 1990, 1098-1104.

[16] S Osowski, 'Neural network for estimation of harmonic components in a power system', *IEE Proceedings*, **139**, 1992, 129-135.

[17] P E Gill, W Murray and M H Wright, *Practical Optimization*, Academic Press, 1981, 133-139.

[18] S V B Aiyer, M Niranjan and F Fallside, 'A theoretical investigation into the performance of the Hopfield model', *IEEE Transactions on Neural Networks*, **1**, 1990, 204-215.

[19] L L Lai, W L Chan, C T Tse and A T P So, 'Real-time frequency and harmonic evaluation using artificial neural networks', *Paper Preprint No PE-273-PWRD-0-01-1997*, Feb 1997, to be published in *IEEE Transactions on Power Delivery*.

15

Artificial Neural Network Applications in Digital Distance Relay

The reach accuracy of distance relays on transmission lines is adversely affected by high fault resistance combined with remote-end infeed that is not measurable at the relaying point. In this chapter, an adaptive setting concept that can overcome this disadvantage is discussed. Optimum relay performance can be achieved without the use of a high speed communication channel. This approach employs the artificial neural network (ANN) concept and is suitable for parallel processing hardware implementation, which is more flexible for widely changing power system conditions.

15.1 INTRODUCTION

In the light of recent developments in digital protection relaying techniques [1], adaptive relaying concepts provide another opportunity for improved performance [2,3]; [2] reports that maximum benefit from adaptive relaying will be achieved from integration with substation control and data acquisition (SCADA) functions and interfacing with the central energy management system (EMS).

Digital distance relays using microprocessor technology have indeed overcome some of the traditional problems in protective relaying. However, even with digital distance relaying, the practice has been to design the scheme on the basis of fixed relay settings. Changes would only be made when the configuration or system was significantly modified. The advantages of adaptive distance protection have been discussed and a practical scheme of adjusting the zone 1 boundary has been proposed [4]. Improved performance was obtained by using an algorithm to adjust the boundary angle. However, if the system conditions vary over a wide range and faults occur through high arc resistance, the relay may lose selectivity.

This chapter presents an adaptive scheme employing the neural network (NN) concept for digital distance relay setting motivated by the desire to respond automatically to the network conditions. The effect caused by changes in power flow is also discussed.

15.2 THE ADVERSE EFFECT OF FAULT RESISTANCE

In conventional distance relays, impedance measurement is made by using the a.c. quantities available at the local end only. It is impossible to determine the infeed current effect through the fault resistance. The remote-end infeed is dependent not only on fault location and fault resistance but also on source impedance of all power sources. The conventional safety margin used in zone 1 is not adequate to avoid mal-operation. A large error results from high fault resistance and sometimes even zone 2 may not be able to provide coverage if the fault resistance is excessive. In EHV transmission lines, a fault resistance up to 200 Ω is possible and therefore a solution to this problem is required.

15.3 AN ADAPTIVE RELAYING SCHEME

System conditions external to the protected line influence relay performance. To demonstrate this effect and outline the adaptive relay setting scheme, a 400 kV three-source system model is used as shown in Figure 15.1. The length of transmission line is 125 km between Local and Remote Systems, 200 km between Local to External, and 200 km between Remote to External. The same line constants are used for all 400 kV lines. A single-line-to-ground fault at F through a fault resistance R_f is examined and a digital distance relay installed at L is considered.

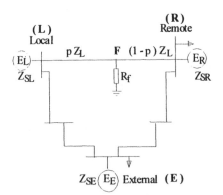

Figure 15.1 Single-line-to-ground fault model

E_L , E_R and E_E are the equivalent potentials at the local, remote and external ends, respectively. Z_{SL} , Z_{SR} and Z_{SE} are the source impedances. Z_L is the line impedance, not referring to the specific sequence component; p is the proportion from the relaying point to the fault and R_f is the fault resistance.

The impedance measured by a distance relay at the local end can be expressed as

$$Z_a = pZ_{1L}\Delta Z \tag{1}$$

where

$$\Delta Z = f_L(Z_{SL}, Z_{SR}, Z_{SE}, Z_L, h_{RL}, h_{EL}, \delta_{RL}, \delta_{EL}, p, R_f) \tag{2}$$

f_L is a non-linear function used to represent Z in terms of the pre-fault and the post-fault parameters [5].

The dependence of Z on pre-fault and post-fault conditions as expressed in Equation (2), introduces a significant difficulty for pilot-independent distance relaying, especially for high resistance faults. The measured impedance may be larger or smaller than the actual value depending on the value of the phasor Z for different power system conditions. Simulation results for different values of R_f and the fault location are given in Figure 15.2. The boundaries can be calculated in the local computer system through SCADA and EMS. Information from the remote and external end can be transferred through a computer link, rather than installing an expensive fast-response data channel exclusively for use during the fault. If a maximum fault-resistance accommodation and relay reach are pre-set, an ideal trip region can be constituted prior to a fault or disturbance using Equations (1) and (2). Setting patterns are renewed (that is, f_L is re-calculated) automatically to accord with changing network conditions.

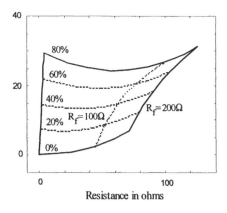

Figure 15.2 Variation of measured impedance with fault resistance for different fault locations

The fault level for all sources is 30 GVA,

$Z_{0SL}/Z_{1SL}=1$, $Z_{0SR}/Z_{1SR}=1$, $Z_{0SE}/Z_{1SE}=1$, $h_{RL}=1$, $\delta_{RL}=-13.5°$, $h_{EL}=1$, $\delta_{EL}=-8.5°$, $Z_{1L}=37.5 \angle 86° \Omega$, $Z_{0L}=136.6 \angle 69° \Omega$, $P_{L\ to\ R}=901$ MW, $Q_{L\ to\ R}=200$ MVAr

15.4 AN IDEAL TRIP CHARACTERISTIC

If system conditions are fixed, R_f and the fault locations are varied, four boundary lines, defined below, can be obtained.

Boundary A: faults at a relay-reach end (80% of line length) with a fault resistance of 0 to 200 Ω
Boundary B: faults at different locations with a 200 Ω fault resistance;
Boundary C: faults at the relaying point with a fault resistance of 0 to 200 Ω
Boundary D: vertical line of constant resistance set at -4 Ω in order to cover solid faults with a zero fault resistance at different locations of the line

The four lines and the included area constitute what may be designated an ideal trip region under the prevailing system conditions (Figure 15.3).

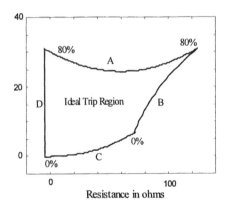

Figure 15.3 An ideal trip region under the same system conditions as in Figure 15.2.

15.5 USE OF ANNS IN IDENTIFYING TRIP REGION

Speed and accuracy are essential in relaying, but to achieve accuracy, a more complex software has to be processed by the indicated accurate mathematical models derived previously. This conflicts with speed requirements. To obtain both speed and accuracy, software pre-fault/during fault realisations have been designed and tested as described below.

15.5.1 Design of an ANN

Three ANNs have been designed for trip region identification for the three non-linear boundaries A, B, and C as indicated in Figure 15.3. Each of them has been tested with a parallel processing approach to adaptive distance relaying [6]. The feedforward network with a backpropagation training function is used. This method tends to give reasonable answers when presented with inputs that ANNs have never seen. Better initial weights and biases are generated using the Nguyen-Widrow initialisation method. The training performance is improved by using an approximation of Newton's method called Levenberg-Marquardt. This optimisation technique is more powerful than the gradient descent method used by the ordinary backpropagation training method but it needs more memory. Because the computers used at present are much more powerful than before in terms of speed and memory, this method proves to be faster and gives more accurate predictions. Each ANN is designed with three layers one input layer with one neuron; one output layer with one neuron; and one hidden layer with a number of neurons. The logistic sigmoid transfer function is used between the input and hidden layers and the pure linear transfer function is used between the hidden and output layers.

For training the ANN for Boundary A, measured values R_A placed at different points in Boundary A are the inputs and, through the hidden layer, the corresponding value X_A is obtained as the output. For training the ANN for Boundary B, measured values X_B at different points in Boundary B are the inputs and, through the hidden layer, the corresponding value R_B is obtained as the output. Boundary C is trained in a manner similar to Boundary A. No training is required for Boundary D as it will remain unchanged under different system configurations [7]. Li et al. [8] have also introduced a better approach in using ΔR and ΔX, but it will not be discussed here.

15.5.2 Pre-fault Setting for On-Line Training and Testing

The ideal boundary for a distance relay, as discussed previously, varies owing to different pre-fault conditions. These conditions, however, can be obtained prior to a fault through the local on-line management computer system. A series of resistance and reactance values corresponding to a different fault resistance for a pre-designed relay reach, say 80% of the line, can then be calculated. Calculated resistances and reactances on the boundary are used as training sets to obtain the weights and biases. The pre-fault training process ends when all weights and biases are converged. This NN constitutes the setting pattern that will be used during faults.

Stringent requirements of reliability for power system relays makes the application of ANN in the adaptive relaying very special in the sense that the trained neural network should be much more reliable in the correct response to faults. Therefore, numerous training and testing patterns should be properly selected to represent different fault measurements [6].

15.5.3   Trip Decision Making

As there are three boundaries, three different NNs are used, one for each boundary. Each network is trained individually until satisfactory results are obtained. With well-trained weights in memory, the NN of each boundary gives an output, Y, according to Equation (3).

$$Y = f_{\text{pure}} \{|W_2| \cdot f_{\log}(|W_1| \cdot |Z|, |B_1|)|, |B_2|\} \tag{3}$$

where $f_{\log}$, f_{pure} are the logistic sigmoid and pure linear transfer functions, respectively, $|W_1|$, $|W_2|$ are the weights of the trained network for layers 1 and 2, respectively, $|B_1|$, $|B_2|$ are the biases of the trained network for layers 1 and 2, respectively, and $|Z|$ is the input to the NN.

Each NN is trained to predict the corresponding boundary. The on-line measurement is fed to each ANN. The output of the ANN is then compared with the on-line measurement to judge whether it is within the trip region. For example, in the ANN for Boundary A, the estimated boundary reactance X_A corresponding to the on-line measured resistance R_M is compared with the on-line measured reactance X_M. If X_A is larger than X_M, then it is inside the trip region. The distance of the measured impedance from Boundary A is then the difference between the ANN output and the measured value. Similar reasoning is applied to Boundaries B, C, and D. On the basis of this approach, the measured impedance that is within the trip region can be identified.

15.5.4   NN Training Result

The number of neurons in the hidden layer affects the accuracy of the non-linear boundary and operating speed. Too few neurons can lead to underfitting, too many can contribute to overfitting, where all training points fit well, but the fitted curve makes wild oscillations between these points. In the present case, two neurons in the hidden layer are found adequate for the ANN for Boundary A and only one neuron in the hidden layer is needed for the ANN for Boundaries B and C. The training time varies according to the starting point and the acceptable error that can be tolerated. As the starting point is chosen at random, in some cases the training may not converge. If so, it is necessary to start again from a different starting point or try with a different number of neurons in the hidden layer. The tolerable sum square error chosen in this study is 1E-7, 1E-6, and 1E-5 for Boundaries A, B, and C respectively. The training converged in 53, 107 and 6 epochs for Boundaries A, B, and C, respectively. The predicted boundary produced by the NN in comparison with the calculated ideal relay reach is shown in Figure 15.4. It can be seen that the NN output matches the ideal relay reach very closely.

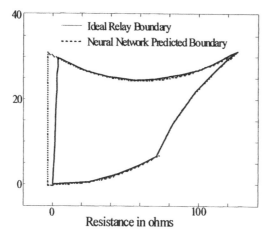

Figure 15.4 Relay boundary

15.5.5 Testing Patterns and Results

Testing patterns should be properly selected to represent different fault measurements. Impedances that are different from the training patterns on the outer and inner curves of boundaries are chosen to test the NN. A typical testing pattern is shown in Figure 15.5.

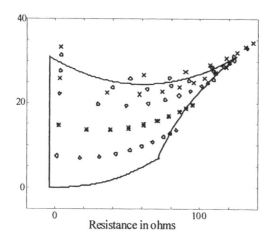

Figure 15.5 Testing pattern

Testing points are generated by simulating single-phase-to-earth faults located at 20%, 40%, 60%, 70%, 75%, 85% and 90% of the line length from the relaying point with fault resistance R_f ranging from 0 to 220 Ω. Both in-zone and out-zone faults at different locations and with different fault resistances are represented. By presenting these testing patterns as inputs to the trained NNs, we can obtain the corresponding outputs by processing Equation (3) for each NN. An internal fault will be identified if the output from all four NNs are positive. The test result for the testing pattern as shown in Figure 15.5 is shown in Table 15.1. Each row corresponds to faults at different locations on the line from the relaying point. The top row shows the corresponding fault resistance.

Table 15.1 Output of NN for testing pattern

R_f	0	25	50	75	100	125	150	180	220 Ω
90%	-3.8	-2.2	-1.6	-1.2	-1.0	-0.8	-0.8	-0.7	-0.7
85%	-1.9	-1.2	-0.9		-0.6		-0.5		
75%	1.9	1.4	1.1	1.0	0.8		0.6	0.6	-1.0
70%	3.7	2.9	2.4	2.0	1.8	1.6	1.4	1.3	-1.3
60%	6.9	6.2	5.3	4.6	4.1	3.7	3.3	2.3	-1.7
40%	6.0	11.9	11.7	10.5	9.6	8.8	8.1	2.8	-2.4
20%	5.0	6.6	6.8	7.3	8.1	8.9	8.8	3.1	-2.9

The result shows that the output is positive if the test points are inside the ideal operating zone and is negative if they are outside the operating zone. The value of the output shows its distance from the nearest boundary in ohms. In some cases, however, if the result is not satisfactory for one or more testing patterns, the NN should be re-trained, including the failed patterns in the training set.

15.6 EFFECT OF POWER FLOW CHANGES ON THE IDEAL OPERATING REGION

15.6.1 Decrease in Power Flow

It is interesting to find out how the ideal operating region is affected by changes of power flow. Figure 15.6 shows how the operating region is affected when the power flow in the line is decreased to 800 MW but with the reactive power flow remaining unchanged. The same system configuration as in Figure 15.1 is used. The ideal operating region should be updated whenever the system configuration or load pattern changes. Since the system load is always changing at a slow rate, one concern about the adaptive method is the possible error introduced if the ideal operating region is not updated for small load changes or if a fault occurs before the updating process is completed. It can be seen from Figure 15.6 that the ideal trip region expands slightly. This means that there is a slight reduction of coverage and the relay will underreach in certain circumstances. The error is higher for high resistance faults at the far end of the line near the pre-set boundary. A similar testing pattern as mentioned in Section 15.5.5 is simulated to test the error caused. The test result is shown in Table 15.2. It can be seen that the reach error is mainly in the line section between slightly under 70% to 80% of the line length from the relaying point and the error is larger for a fault resistance above 100 Ω.

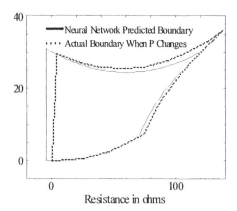

Figure 15.6   Changes in relay coverage owing to decrease in the power flow of the line

Table 15.2   Output of NN for decrease in power flow

R_f	0	25	50	75	100	125	150	180	220 Ω
90%	-3.8	-3.4	-3.3	-3.1	-3.0	-3.1	-3.2	-3.4	-3.7
85%	-1.9	-2.3	-2.4		-2.5		-2.6		
75%	1.8	0.7	-0.02	-0.5	-0.8		-1.2	-1.4	-1.6
70%	3.7	2.3	1.4	0.8	0.3	-0.03	-0.3	-0.5	-0.8
60%	6.9	5.7	4.5	3.6	2.9	2.3	1.9	1.4	-0.5
40%	6.0	12.0	11.2	9.9	8.8	7.9	6.5	1.1	-4.0
20%	5.0	6.6	6.9	7.6	8.4	9.3	9.7	0.6	-5.7

15.6.2   Increase in Power Flow

On the other hand, when the power flow increases to 1000 MW while the reactive power flow remains approximately constant, there is a slight increase in coverage and again the error is higher for high resistance faults. It can be seen from Figure. 15.7 that the relay will overreach for a fault resistance higher than 50 Ω. A similar testing pattern is used to test the resultant error. The test result is shown in Table 15.3.

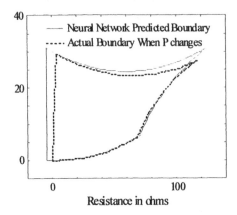

Figure 15.7   Changes in relay coverage owing to increase in the power flow of the line

Table 15.3   Output of NN for increase in power flow

R_f	0	25	50	75	100	125	150	180	220 Ω
90%	-3.7	-1.1	-0.04	0.5	0.9	-0.5	-3.0	-4.9	-6.5
85%	-1.8	-0.3	-0.4		-1.1		-0.5		
75%	1.9	2.1	2.2	2.3	2.3		2.3	-0.1	-3.4
70%	3.8	3.5	3.4	3.2	3.1	3.0	3.0	1.0	-2.7
60%	6.9	6.6	6.0	5.6	5.2	4.9	4.7	2.8	-1.6
40%	6.0	11.7	11.5	11.1	10.3	9.7	9.1	4.5	-0.7
20%	5.0	6.5	6.6	7.1	7.7	8.4	9.2	5.2	-0.5

15.6.3   Increase in Active and Reactive Power Flows

When both the active and reactive power flows in the line increase to 951 MW and 225 MVAr respectively, it can be seen from Figure 15.8 that there is a slight underreach for resistance faults above 100 Ω near the setting limit in Boundary A. A similar testing pattern is generated to test the resultant error. The test result is shown in Table 15.4.

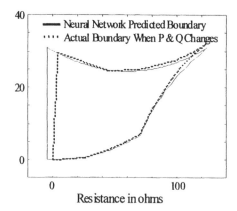

Figure 15.8 Changes in relay coverage owing to increase in the active and reactive power flows of the line

Table 15.4 Output of NN for increase in active and reactive power flows

R_f	0	25	50	100	150	180	220 Ω
90%		-2.3	-2.1	-2.3		-2.6	
85%		-1.3	-1.3	-1.6		-2.1	
75%		1.4	1.0	0.1	-0.4	-0.7	-1.0
70%	3.8	3.0	2.3	1.2	0.5	0.1	-0.2
60%	6.9		5.2	3.7	2.6	2.1	1.6
50%	6.4		8.3	6.4		4.3	1.8
25%	5.3		8.2	9.5	10.2	5.0	1.3

15.7 PROPOSED TRIPPING SCHEME

The output from the ANN indicates the distance of the measured impedance from the nearest boundary. In the case study, it can be shown that if this value is less than 1 Ω, then the fault is very close to the boundary. If the power flow prior to the fault is larger than the trained condition for 100 MW or more, then the relay is most likely to have overreach problems. In this case, tripping should not be initiated immediately if the NN measured impedance from the boundary is less than 1 Ω. Instead, a time delay should be initiated to allow the updating or renewal of the pre-stored weights and biases of the ANN for the nearest loading condition in order to allow for a more precise measurement to be made. If the power flow prior to the fault is less than the trained condition for 100 MW or more, then the relay is most likely to have underreach problems. A change in reactive power would also cause similar effects. It is also noted that if the difference between the power flow prior to the fault and the trained condition is not more than 100 MW, and the NN measurement from the boundary is more than 2 Ω, the fault is well within the trip region and tripping should be triggered without delay. If the NN measurement from the boundary is less than 2 Ω, then a time delay that is inversely proportional to the measured impedance can be adopted to avoid unnecessary tripping.

15.8 OFF-LINE TRAINING OF THE NN

From the previous discussions it can be seen that a change of the system operating condition can give rise to a different ideal trip region. To cater for the training requirements for each condition, families of hypothetical pre-fault conditions and during-fault 'measurements' have to be supplied in advance. In the present case study, pre-trained NN should be available for every 100 MW interval. The relay can then be trained initially, off-line, before it is put into use. The most appropriate pattern of trained weights and biases can then be selected depending upon the system configuration and conditions.

15.9 CONCLUSIONS

The concern for the time being is with economics, owing to the requirement of high performance microprocessor chips. However, by virtue of the fact that electronic engineering is rapidly developing and analog VLSI techniques for NNs are available [9], the cost of implementation will decline in the future. If the price becomes acceptable, this approach could practically improve the performance of protective relaying in power systems and application-specific IC chips should be considered for these applications.

Adaptive distance relaying solutions to high fault resistance combined with pre-fault system conditions has been presented. An improved performance of digital distance relays integrated with an intelligent technique has been established. The scheme is designed to operate in conjunction with the existing protection systems. Disabling the adaptive functions does not degrade the non-adaptive functions of the relays.

15.10 REFERENCES

[1] M S Sachdev (Co-ordinator), *Microprocessor relays and protection systems*, IEEE Tutorial Course Text No. 88 EH0269-1-PWR, Piscataway, NJ, 1988.
[2] G D Rockefeller, C L Wagner and J R Linders, 'Adaptive transmission relaying concepts for improved performance', *IEEE Transactions on Power Delivery*, **3**, 1988, 1446-1458.
[3] IEEE Power System Relaying Committee, 'Feasibility of adaptive protection and control', *IEEE Transactions on Power Delivery*, **8**, 1993, 975-983.
[4] Z Zhang, and D Chen, 'An adaptive approach in digital distance protection', *IEEE Transactions on Power Delivery*, **6**, 1991, 135-142.
[5] Y Q Xia, K K Li and A K David, 'Adaptive relay setting for stand-alone digital distance protection', *IEEE Transactions on Power Delivery*, **9**, 1994, 480-491.

[6] K K Li, A K David, L L Lai and Y Q Xia, 'An adaptive digital distance relay for EHV transmission lines', *IFAC-95 Symposium on Large Scale Systems*, **1**, London, 1995, 433-438.

[7] K K Li, L L Lai and A K David, 'Intelligent digital distance relay', *Proceedings of the International Conference on Intelligent System Application to Power Systems*, The Korean Institute of Electrical Engineers, July 1997, 279-283.

[8] K K Li, L L Lai and A K David, 'Stand alone intelligent digital distance relay', to be published in *IEEE Transactions on Power Systems*.

[9] E M Georges, L L Lai, F Ndeh-Che, and H Braun, 'Neural networks implementation with VLSI', *Proceedings of the Fourth International Conference on Artificial Neural Networks*, IEE, 1995, 489-494.

16

Application of Artificial Neural Networks to Transient Stability Assessment

This chapter presents an evaluation of the effectiveness of artificial neural networks (ANNs) for the rapid determination of critical clearing times for power networks with varying line outages and load patterns. Studies are reported on the performance of ANNs that have been trained using previously proposed and new training items. These show that for small power networks, good results could be obtained. However, for large networks there is a difficulty in returning consistently accurate answers under varying network conditions.

16.1 INTRODUCTION

Despite major research efforts, the literature suggests that there have been very few practical applications for ANNs in the field of power engineering. The situation with respect to practical applications of ANNs contrasts with the adoption of other artificial intelligence techniques such as expert systems [1] and fuzzy technology [2]. This contrast could be a due to the current imperfect state of knowledge of ANNs, despite their long history, or it could be due to attempts to apply ANNs to inappropriate problems.

This chapter reports on the effectiveness of ANNs for the on-line determination of the transient stability of electrical power systems. Investigations have included proposed techniques [3] and enhancements, which are discussed here. After exhaustive testing of the techniques on practical systems [4], it is concluded that at present there are inherent limitations of ANNs that inhibit their use in this application.

The backpropagation neural networks could be useful when applying to test patterns that have a high correlation with the training patterns. As, in general, this NN is not self-learning, it is to accommodate relatively minor deviations in power system operating conditions and topology.

16.2 TRANSIENT STABILITY

There are mainly three approaches to the transient stability assessment problem. The conventional numerical integration method discretises the machine swing equations to obtain the evolution with time of the machine rotor angles. This method is tedious and not suitable for an on-line situation [5-9]. The pattern recognition method is based on developing on-line tools for transient stability assessment depending on information obtained off-line [10, 11]. The difficulty in the selection of features and development of the learning set have limited its application in an on-line environment. For the direct method [6], this involves the application of the Lyapunov stability theory of non-linear systems to the transient stability problem. Even though this approach is accurate, it is computationally tedious and thus it is not suitable for on-line applications. For high-speed assessment, the direct method of Lyapunov explores the stability of non-linear differential equations using techniques developed by Lyapunov in the 1890s. In this approach, an attempt is made to establish a region within which the system is stable. Several problems arise in the application of the Lyapunov method, and to date it appears to have had only limited practical applications. A major problem is the choice of an appropriate Laypunov function [3]. Apart from the above methods, fast assessment of transient stability limits of power systems based on ANNs can be an alternative method. The ANN is capable of mapping the pre-fault and post-fault system features into the critical clearing time in negligible time [4, 12-14].

The degree of modelling required for stability studies depends largely on the purposes for which the proposed stability studies are required [15]. Transient stability behaviour could not be totally understood by considering pre-fault or static variables. A dynamic variable such as the initial machine acceleration is important as this reflects the machine dynamics, the changes in the network topology, fault locations, initial operation conditions and fault types. Transient stability depends on the total system load level. For any load level, transient stability behaviour differs with different distributions of loads and generations. Transient stability also depends on the severity of disturbance.

16.3 CRITICAL CLEARING TIME (CCT)

The type of disturbance which is most important in stability studies is a fault applied and subsequently cleared. It is usually desired to determine whether a system is stable with a given load and fault-clearing time. The concept of a critical clearing time is well known [15], it represents the maximum time that a particular fault can be allowed to persist in a system without instability occurring. The most consistently reliable method for obtaining the CCT is to apply a fault and to analyse the system behaviour in the time domain. this method of obtaining the CCT is very time consuming and is most suitable for off-line design or planning work rather than for on-line operational purposes.

In this chapter, results are obtained with a system model based on a number of commonly used assumptions. Generators are represented as constant voltages behind transient reactance, turbine output powers are assumed constant and loads are modelled by constant impedance.

16.4   METHODS OF FAST ASSESSMENT OF CCT

The capabilities of the ANN to learn and generalise enable the network to obtain the complex mapping that carries pre-fault and post-fault system attributes on to the single valued space of critical clearing time. The CCT is an attribute that provides important information about the quality of post-fault system behaviour and it can be estimated by the ANN in negligible time. Recently, ANN approaches have been proposed [3, 4, 12-14] for the estimation of CCTs in transient stability assessments of power systems.

Application of the ANN to the determination of the CCT removes the necessity for on-line time domain solutions since the trained ANN is simply presented with data on the current system operating state and the CCT is provided as output.

As usual, to ensure adequate training of an ANN, it is necessary to use a sufficiently large number of CCTs assembled with other information into sets of training facts. A useful ANN should be trained under a range of different system loading conditions and pre-existing line-out conditions.

16.5   NEURAL NETWORK (NN) APPLICATION

In the case of the backpropagation ANN, there will always be an input layer and an output layer and one or more hidden layers. Each item of input data will have an associated input neuron and each item of output data will have an associated output neuron. In the case of first swing stability, the only output is the CCT.

In this chapter, different kinds of input features are discussed. The practical system considered is more complex and it can show the limitations and capabilities of a NN application in the area.

16.5.1   Training Patterns

A number of patterns related to the system conditions are assembled as training sets to ensure adequate training is presented to the ANN during the training phase. The selection of the training features used to form the training pattern is of critical importance to the success of operating an ANN.

Since the intention is to provide fast, on-line assessment of the CCT, it is necessary to ensure that the input data required for obtaining an answer from the

ANN is readily available. Steady-state power system network solutions can be obtained very rapidly, and provide a good source of information for ANN training.

16.5.2 Training Features

Training features comprise useful information obtained from a series of steady-state network solutions in the intact, faulted and fault-cleared states. The CCT obtained from dynamic time-domain analysis is also included as a training feature.

For steady-state solutions, the network is reduced to the internal machine buses and their equivalent interconnections. The internal voltages and mechanical power inputs on the individual machines are held constant. The features comprising a training pattern are then chosen as follows [4]:

- **Feature 1:** Accelerating power for each machine immediately following the application of a fault

$$P_{ai} = P_{mi} - P_{ei} \tag{1}$$

where P_{ai} is the accelerating power available for machine i; P_{mi} is the mechanical input power for machine i (mechanical power is equal to the initial steady state electrical load on the machine plus machine losses); P_{ei} is the electrical output power for machine i (electrical power output is calculated under the specified network conditions).

Dividing the accelerating power by the machine inertia M_i as proposed in [3], is equivalent to scaling. Since this feature in all training patterns is modified in the same way, no improvement in the performance of the ANN is observed

- **Feature 2:** Accelerating power is squared for each machine immediately following the application of a fault. This feature was scaled by dividing by the machine inertia and is claimed to give information on individual machine accelerating energy.

- **Feature 3:** Rotor angle, δ_{di}, for each machine relative to the reference machine, with the faulty line removed, where δ_{di} is equal to the angular deviation of rotor i with respect to the reference machine.

- **Feature 4:** Driving point susceptance, B_{ii}, which represents the post-fault strength of the connection of the machine to the rest of the network.

- **Feature 5:** Total system 'energy adjustments'

$$EN = \sum_{i=1}^{n} \left(P_{ai} \cdot \delta_{di} \right) \tag{2}$$

where *EN* is defined as the energy adjustments in the system and *n* is the number of machines in the study.

• **Feature 6:** Critical Clearing Time
The maximum set of training features for each training pattern therefore comprises four features per machine and two features relating to the system (Table 16.1).

Table 16.1   Features used in each training pattern

Machine Features	P_{ai} fault on	P_{ai}^2 / M_i fault on	δ_{di} post-fault	B_{ii} post-fault
System Features	*EN*	CCT		

The training features were tried in all possible combinations from the use of two features per training pattern up to the maximum of six, one of which was always the CCT.

It is important to ensure that the range of training data in the training pattern presented to the ANN is adequate for the problem being posed. ANN may be thought of as a very good interpolator but its powers of extrapolation are limited, if indeed they exist at all. It is essential that interpretation of the NN is within the range of the set of training patterns on which the network has been trained.

16.5.3   Pre-Conditioning of Training Patterns

When the training features are first assembled as training patterns they may not be in the most effective form for training. Pre-conditioning of the training data, such as scaling, is generally necessary. Various ways of pre-conditioning data were used to improve training, including dividing each feature of the raw data by the maximum value in the set dividing through by the average and dividing through by the variance. This is an interesting area for future research since training was found to be significantly influenced by pre-conditioning.

When training a NN, it is a matter of chance whether a satisfactory ANN is obtained. Should the NN not train satisfactorily, or show signs of becoming unstable after a few thousand training cycles, it may be necessary to randomise the weights or to try a variation of the pre-conditioning and hope that the next training solution is more satisfactory.

16.6   PERFORMANCE OF THE ANN

Studies were performed on two practical power systems [4] as shown in Table 16.2.

Table 16.2 System parameters (Table II from [4], reproduced by permission by the IEEE)

	System 1	System 2
Total load	200 MW	7000 MW
Generating stations	4	20
Buses	12	106
Lines	16	156
ANN		
Number of layers	3	3
Number of input neurons	17	81
Number of hidden neurons	17	68
Number of output neurons	1	1

A four-layer ANN with various numbers of neurons in the two hidden layers was tried on the system but there appeared to be little improvement in performance. Increasing the number of neurons in the single hidden layer of the three layer ANN defined above did not appear to improve performance either [4].

16.6.1 Training

To obtain training patterns, the system was solved at five load levels and generation levels. In addition, solutions were obtained at each of the five load levels under three different topological conditions. In the first condition, all lines were in service. In the second condition, a significant line was removed. In the third condition, a different significant line was removed.

For a number of different system operating conditions, a fault was applied to a bus to simulate a line fault close to that bus. The faulted line was then removed and the CCT for this location and condition was obtained. This process was repeated until every line in the system had been faulted in turn at each end. When all the training patterns had been assembled, the ANN was subject to 20000 training cycles.

Extensive testing was carried out to determine the relative training effectiveness of the various features. It appeared that none of the features was effective individually for training the NN, but that all contributed to some extent in producing a trained NN when used together.

The training performance of the ANN was assessed by examining the maximum per-unit error of the presented CCTs. Assuming that if the maximum error was less than 0.1 per unit, then the ANN was considered to be satisfactorily trained. The performance of the network was acceptable for some load patterns but not for others. When more variable conditions were included, the error occasionally rose to an unacceptable level of up to 0.5 per-unit.

For a larger system, additional difficulties were encountered. Much more time was required to calculate the many CCTs needed for the training patterns and the times for training the ANN also increased sharply. Although a number of attempts were made to apply ANN techniques to this system, it proved not to be easy.

16.6.2 Testing

The trained ANN, when presented with previously unseen patterns, produces a series of CCTs that ideally should closely approximate the CCTs determined by the traditional time domain method.

In the course of this investigation, many of the presented patterns did result in very low errors, but there were also many that resulted in unacceptably large errors.

16.7 DISCUSSION

Considerable difficulties were experienced in training the ANN. The ANN could produce good results towards the middle of the CCT training range but values at the extremities tended to become unacceptable. Various attempts have been tried to emphasise the end values. The obvious method of increasing the number of training patterns present at the extremities did not have the desired effect and this remains a problem.

Increasing the training from 20000 to 200000 cycles did not improve the performance significantly. The technique of polling a number of ANNs trained on differently randomised untrained networks also failed to improve performance.

As could be expected, the ANN generally trained better when all data came from a network with conditions that varied very little. Under these conditions the maximum training error of a collection of data could easily be below 0.1 per unit.

An ANN that is generally well trained will nevertheless occasionally produce erratic and erroneous results when testing with new patterns. This unpredictable behaviour would generally be unacceptable in practice.

Despite extensive training using many different sets of training patterns and various methods of pre-conditioning, it was not possible to obtain a satisfactorily trained ANN under significantly varying network conditions.

The reason for the poor performance of the ANN on this particular problem is open for discussion. It seems likely that the difficulty stems from widely varying network conditions and fault locations leading to very similar CCTs, whereas quite different CCTs can arise from relatively small variations in network conditions, such as choice of fault locations. Under such circumstances, no pattern may be found, even by an ANN.

However, the prediction of using the CCT with the ANN could produce good results for a small power system. The ANN has been applied to a sample single machine system and the estimation of the CCT is excellent [14]. In this case, rotor angle, fault distance, terminal voltage and current after the fault inception are used as the input features. In [13], the ANN has been applied to the estimation of the CCT. The active and reactive power of each generator are selected as the input features to form the input patterns for the ANN. The system consists of three machines and nine busbars. The maximum prediction error on the CCT is 8.3%.

16.8 CONCLUSIONS

The range of CCT errors from the ANN on testing by previously unseen patterns, and in some cases patterns that have been used in the training process, is at present too large for the use of this method on a practical system. However, it is believed that with the advancement of computing facilities, an ANN approach is possible in the near future.

The success of an ANN approach for power system transient stability assessment depends upon the successful learning of the correct mapping by the ANN. This, in turn, depends on numerous factors such as the selection of input features, the scaling of these features and the selection of training patterns and neural network architectures.

16.9 REFERENCES

[1]   Z Z Zhang, G S Hope and O P Malik, 'Expert systems in electrical power systems - a bibliographical survey', *IEEE Transactions on Power Systems*, **4**, 1989, 1355-1362.

[2]   D G Schwartz, 'Fuzzy logic flowers in Japan', *IEEE Spectrum*, **29**, 1992, 32-35.

[3]   D J Sobajic and Y H Pao, 'Artificial neural-net based dynamic security assessment for electric power systems', *IEEE Transactions on Power Systems*, **4**, 1989, 220-228

[4]   E Hobson and G N Allen, 'Effectiveness of artificial neural networks for first swing stability determination of practical system', *IEEE Transactions on Power Systems*, **9**, 1994, 1062-1068.

[5]   B Toumi, R Dhifaoui, ThVan Cutsem and Ribbens-Pavella, 'Fast transient stability assessment revisited', *IEEE Transactions on Power Systems*, **1**, 1986, 211-220.

[6]   M Ribbens-Pavella and F J Evans, 'Direct methods for studying dynamics of large-scale electric power systems', *Automatica*, **21**, 1985, 1-21.

[7]   O Gurel and L Lapidus, *A Guide to Methods for the Generation of Lapunov Functions*, International Business Machines Corporation, New York, 1968.

[8]   M A Pai, *Energy Function Analysis of Transient Stability*, Kluwer, 1989.

[9]   H W Dommel and N Sato, 'Fast transient stability solutions', *IEEE Transactions on Power Apparatus and Systems*, **91**, 1972, 1643-1650.

[10] C S Chang, D Srinivasan and A C Liew, 'A hybrid model for transient stability evaluation of interconnected longitudinal power systems using neural/pattern recognition approach', IEEE Power Engineering Society, *Paper No 93 WM 153-7 PWRS*.

[11] F S Prabhakara and C T Heydt, 'Review of pattern recognition methods for rapid analysis of transient stability', *THO169-3/8710000-0016*, IEEE, 1987.

[12] K Huang, D Lam and H Yee, 'Neural-net based critical clearing time prediction in power system transient stability analysis', *Proceedings of the Second International Conference on Advances in Power System Control, Operation and Management*, IEE, Dec 1993, 679-683.

[13] K L Lo and R J Y Tsai, 'Power system transient stability analysis by using modified Kohonen network', *Proceedings of the International Confernece on Neural Networks*, IEEE, Australia, 1995.

[14] B S Lau and K P Wong, 'An artificial neural network approach to transient stability assessment', *Australian Journal of Intelligent Information Processing Systems*, Autumn 1996, 75-85.

[15] B M Weedy, *Electric Power Systems*, John Wiley & Sons, 1979.

17

Application of Neural Networks and Evolutionary Programming to Short-Term Load Forecasting

The feedforward neural network (FNN) and evolutionary programming (EP) for short-term load forecasting are presented in this chapter. A load forecasting model has been developed using a multilayer perceptron NN with an appropriately modified backpropagation learning algorithm. The model produces a short-term forecast of the load in the 24 hours of the forecast day concerned. The technique has been tested on data provided by an Italian power company and the promising results obtained through the application of object-oriented programming neural network (OOPNN) based approach show that the approach is very effective in that it gives accurate predictions. For the EP-ANN, the approach could also produce similar results. However, for some poor initial starting points, because the problem is very complex, it takes a much longer time to obtain a comparable solution. This points in the direction of evolutionary computing being integrated with parallel processing techniques to solve practical problems.

17.1 INTRODUCTION

An accurate and stable load forecast is essential for many operating decisions taken by utilities. In fact, it is well known that a cheap and reliable power system operation is definitely the result of good short-term load forecasting. The short-term load forecast provides the information to be adopted in the on-line scheduling and security functions of the energy management system, such as unit commitment, economic dispatch and load management. Hence, accurate load forecasting is essential for the optimal planning and operation of large-scale power systems.

Many techniques have been proposed and used for short-term load forecasting. Time-series models based on extrapolation are used for the representation of load behaviour by trend curves. The time series approach, regression approach, state-space models, pattern recognition and expert systems are also some of the other techniques used [1-5].

The time series approach assumes that the load of a time depends mainly on previous load patterns, such as the auto-regressive moving average models and the spectral expansion technique [2].

The regression method utilises the tendency that the load pattern has a strong correlation to the weather pattern. The weather-sensitive portion of the load is arbitrarily extracted and modelled by a pre-determined functional relationship with weather variables.

All the above approaches use a large number of complex equations that involve much computational time.

More recently, artificial neural network (ANN) techniques have been used by several researchers [6-21]. These techniques have poor convergence during training.

In this chapter, the OOP technique is used to design a NN. If many NNs are required, then the inheritance properties could be used to reduce the time to re-design new NNs [22]. A NN can be specified at the highest level in terms of architectures, motivation function, learning and update rules. Also, as from current literature [20], it shows that EP-ANN has some advantages over the ANN with a BP learning approach. This chapter shows that the benefit derived from EP-ANN is problem-dependent.

17.2 LOAD FORECASTING WITH ANNS

Power load demand is sensitive to weather variables, such as temperature, wind speed, relative humidity and cloud cover (sky cover/light intensity). Although the daily load profile depends on such weather variables, only temperature is considered here.

There are 58 inputs to the developed object-oriented multilayered perceptron (OO-MLP) ANN. The features that are taken into account as input factors in the load forecast system are as follows: two 24-hour load records of day i-1 and i-2 (the forecast day is day i). Six more inputs are the maximum and minimum temperatures of day i, i-1 and i-2. Three inputs as a binary code to show seven days of the week. One binary code is dedicated to the holidays or any yearly special occasions that may effect the forecast. In summary, the designed NN is of the multilayered preceptron type and is used to learn about the relationship between the 58 inputs and 24 outputs.

The inputs are:

- Hourly loads for two days prior to the forecast day   24
- Hourly loads for the day prior to the forecast day   24

- Max. and Min. temps for two days prior to the forecast day   4
- Max. and Min. temps for the forecast day   2
- Day of the week   3 bits
- Holiday   1 bit

The outputs are:

- Load forecast for all 24 hours of the day   24

The network is shown in Figure 17.1. The above values are normalised as indicated by Equation (1).

$$\text{Normalised} = \frac{\text{Actual Value - Min.}}{\text{Max. - Min.}} \tag{1}$$

where Min. and Max. are the maximum and minimum of the attribute, respectively.

The mean square error (MSE) is used to measure the accuracy of the OO-MLP. The sigmoid activation function is adopted.

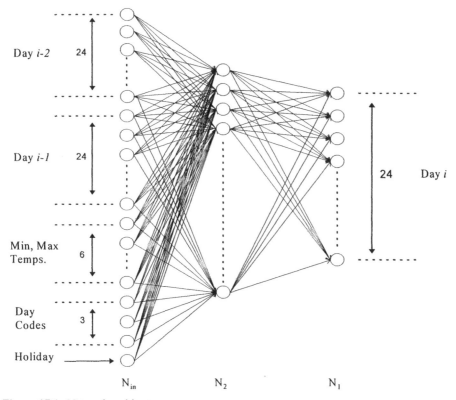

Figure 17.1 Network architecture

The number of neurons in the hidden layer is determined by the trial-and-error method using the validation set. This guarantees the proper selection of the number of neurons in the hidden layer. Various neurons have been tested and the MSE error on training, validation and testing has been measured. With 37 neurons in the hidden layer, it produces the lowest error on the validation set.

17.3   SIMULATION RESULTS

17.3.1   The NN Approach

From the practical data, three training, validation and testing sets are selected using the cross-validation technique. The total number of training patterns is 90. Each pattern represents the loads for a single day. Out of the three months' of data available, 90 patterns are created and 83 used for training and validation. The remaining seven patterns are used to test the OO-MLP ANN. The testing set is unseen by the trained system. Figure 17.2 shows the results from the network. The actual load values and forecast values are shown together in the same figure.

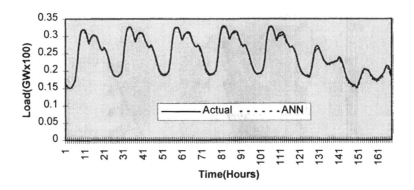

Figure 17.2   Comparison between practical data and neural network results

Table 17.1 shows the final MSE after 15000 iterations for 83 patterns. The average error of the test data is obtained by using Equation (2) as shown below:

Table 17.1 MSE and average error from NNs

Training and Validation sets Final MSE for 83 days after 15000 Iterations	6.54327E-07
Testing Set Average Error for 7 days	1.0362 %

$$\text{Test error } = \left| \frac{\text{Actual load - ANN output}}{\text{Actual load}} \right| \text{ x } 100\% \qquad (2)$$

Figure 17.3 shows the MSE for training between the first 4000 iterations.

MSE of Training Set

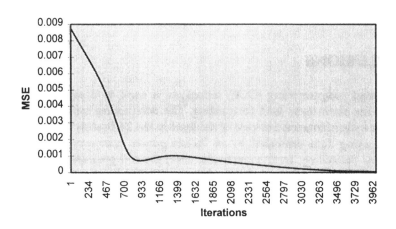

Figure 17.3 MSE vs. number of iterations

17.3.2 The EP-ANN Approach

An EP-ANN designed with the approach as mentioned in Chapter 3 is also applied to the same load forecast problem. A comparison between the results from the ANN and EP-ANN is shown in Figure 17.4. For the EP-ANN case, the number of parameters is 3095. With 4 bytes to represent each parameter, each individual will take up 12180 bytes. With a population size of 100, the memory size is 1218000 bytes. The required computational time is much greater than that for the OO-MLP ANN.

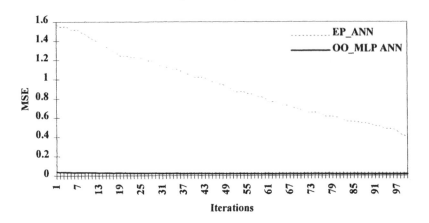

Figure 17.4   Comparison between EP-ANN and OO-MLP ANN

17.4 CONCLUSIONS

The object-oriented programming (OOP) technique is used to design a suitable neural network for short-term load forecasting. The forecasting model has been used to produce a simultaneous forecast of the load in the 24 hours of the forecast day concerned, using data provided by an Italian power company. The results obtained are very promising. In this particular case, the comparison between the results from the EP-ANN and NN shows that the EP-ANN does not provide a faster solution than the NN. This is owing to the fact that the initial randomly selected starting solution is a poor one. The size of the problem is very large and as such the amount of memory and computation time is large also. This points in the direction of parallel processing techniques being integrated with evolutionary computing to solve complex practical problems.

17.5 REFERENCES

[1]   G Gross and F D Galiana, 'Short term load forecasting', *Proceedings of IEEE*, **75**, 1987, 1558-1573.

[2]   M T Hagan and S M Behr, 'The time series approach to short term load forecasting', *IEEE Transactions on Power Systems*, **2**, 1987, 785 - 791.

[3]   A D Papalexopoulos and T C Hesterberg, 'A regression-based approach to short-term load forecasting', *IEEE Transactions on Power Systems*, **5**, 1990, 1535-1547.

[4]   S Rahman and R Bhantnagar, 'An expert system based algorithm for short-term load forecast', *IEEE Transactions on Power Systems*, **3**, 1988, 392-399.

[5]   A S Dhdashti, J R Tudor and M C Smith, 'Forecasting of hourly load by pattern recognition: a deterministic approach', *IEEE Transactions on Power Apparatus and Systems*, **101**, 1982, 900-910.

[6] K Y Lee and J H Park, 'Short-term load forecasting using an artificial neural network', *IEEE Transactions on Power Systems*, **7**, 1992, 124 - 132.

[7] M.Djukanovic, B Babic, D J Sobajic, and Y H Pao, 'Unsupervised/supervised learning concept for 24-hour load forecasting', *IEE Proceedings-C*, **140**, July 1993, 311-318.

[8] M Caciotta, R Lamedica, V Orsolini Cencelli, A Prudenzi and M Sforna, 'Application of artificial neural networks to historical data analysis for short-term electric load forecasting', *European Transactions on Electrical Power*, **7**, 1997, 49-56.

[9] D C Park, M A El-Sharkawi, R J Marks II, L E Atlas and M J Damborg, 'Electric load forecasting using an artificial neural network', *IEEE Transactions on Power Systems*, **6**, 1991, 442-449.

[10] T M Peng, N F Hubele and G G Karady, 'Advancement in the application of neural networks for short term load forecasting', *IEEE Transactions. on Power Systems*, **7**, 1992, 250-258.

[11] O Mohammed, D Park, R Merchant, T Dinh, C Tong, A Azeem, J Farah and C Drake, 'Practical experiences with an adaptive neural network short term load forecasting system', *IEEE Transactions on Power Systems*, **10**, 1995, 254-265.

[12] A G Bakirtzis, V Petridis, S J Kiartzis, M C Alexiadis and A H Maissis, 'A neural network Short Term load forecasting model for the Greek power system', *IEEE Transactions on Power Systems*, **11**, 1996, 858-863.

[13] A Khotanzad, Rey-Chue Hwang, A Abaye and D Maratukulam, 'An adaptive modular ANN hourly load forecaster and its implementation at electric utilities', *IEEE Transactions on Power Systems*, **10**, 1995, 1716-1722.

[14] A Piras, A Germond, B Buchenel, K Imhof and Y Jaccard, 'Heterogeneous artificial neural network for short term electrical load forecasting', *IEEE Transactions on Power Systems*, **11**, 1996, 397-402.

[15] B Kermanshahi, M Zhao, RYokoyama and M Asari, 'A sense of security for mid-term load forecasting using neural net', *IEEE/KTH Stockholm Power Tech International Symposium on Electric Power Engineering*, 1995, 576-581.

[16] R Lamedica, A Prudenzi , M Sforna, M Caciotta and V Orsilini Cencelli, 'A neural network based technique for short term load forecasting of anomalous load periods', *IEEE Transactions on Power Systems*, **7**, 1996, 185-189.

[17] B Kermanshahi, R Yokoyama and K Takahashai, 'Practical implementation of neural nets for weather-dependent load forecasting and re-forecasting at a power utility', *Proceedings of twelfth Power Systems Computation Conference*, PSCC, 1996, 217-223.

[18] S Rahman and O Hazim, 'A generalised knowledge based short term load forecasting system', *IEEE Transactions on Power Systems*, **8**, 1993, 508-512.

[19] T Maifield and G Sheble, 'Short term load forecasting by neural network and a refined genetic algorithm', *Electrical Power Systems Research*, **31**, 9-14.

[20] L L Lai, A G Sichanie, N Rajkumar, E Styvaktakis, M Sforna and M Caciotta, 'Practical application of object oriented techniques to designing neural networks for short-term electric load forecasting', *Proceedings of the Energy Management and Power Delivery Conference*, IEEE, Singapore, March 1998, 559-563.

[21] S E Papadakis, J B Theocharis, S J Kiartzis and A G Bakirtzis, 'A novel approach to short-term load forecasting using fuzzy neural networks', *IEEE Transactions on Power Systems*, **13**, 1998, 480-492.

[22] F N Che, Object oriented analysis and design of computational intelligence systems, *PhD thesis*, City University, London, UK, 1996.

SELECT BIBLIOGRAPHY

K K Li, 'Intelligent adaptive digital distance relaying for high relaying for high resistance earth faults', *PhD thesis*, City University, London, UK, 1998.

K P Wong and J Yuryevich, 'Evolutionary programming based algorithm for environmentally constrained economic dispatch', *IEEE Transactions on Power Systems,* **13**, 1998, 301-306.

R A Gallego, A Monticelli and R Romero, 'Transmission system expansion planning by an extended genetic algorithm', *IEE Proceedings - Generation, Transmission and Distribution,* **145**, 1998, 329-335.

Y M Park, J R Won, J B Park and D G Kim, 'Generation expansion planning based on an advanced evolutionary programming', *IEEE Power Engineering Society,* Paper Preprint Number PE-304-PWRS-0-2-1998.

K Y Lee, A Sode-Yome and J H Park, 'Adaptive Hopfield neural networks for econmoic load dispatch', *IEEE Transactions on Power Systems,* **13**, 1998, 519-526.

A Berizzi, P Finazzi, D Dosi, P Marannino and S Corsi, 'First and second order methods for voltage collapse assessment and security enhancement', *IEEE Transactions on Power Systems,* **13**, 1998, 543-551.

I F Mcgill and R J Kaye, 'Decentralized coordination of power system operation on using dual evolutionary programming', *IEEE Power Engineering Society,* Paper Preprint Number PE-44-PWRS-0-12-1997.

K Warwick, R Aggarwal and A Ekwue, Eds, *Artificial Intelligence Techniques in Power Systems,* IEE, UK, 1997.

K Y Lee and F F Yang, 'Optimal reactive power planning using evolutionary algorithms: a comparative study for evolutionary programming, evolutionary strategy, genetic algorithm, and linear programming', *IEEE Power Engineering Society,* Paper Preprint Number PE-958-PWRS-1-04-1997.

J H Kim and H Myung, 'Evolutionary programming techniques for constrained optimization problems', *IEEE Transactions on Evolutionary Computation,* **1**, 1997, 129-140.

T Bäck, D B Fogel and Z Michalewicz, Eds, *Handbook of Evolutionary Computation,* Oxford University Press and Institute of Physics, New York, 1997.

L L Lai, W L Chan and A T P So, 'A two-ANN approach to frequency and harmonic evaluation', *Proceedings of the Fifth International Conference on Artificial Neural Networks,* IEE, July 1997, 245-250.

A G Jongieper and L Van Sluis, 'Adaptive distance protection of double-circuit lines using artificial neural networks', *IEEE Transactions on Power Delivery,* **12**, Jan 97, 97-105.

R Romero, R A Gallego and A Monticelli, 'Transmission system expansion planning by simulated annealing', *IEEE Transactions on Power Systems,* **11**, 1996, 364-369.

IEEE/Power engineering Society, *A Tutorial Course on Artificial Neural Networks With Applications to Power Systems,* M. El-Sharkawi and D. Niebur, Editors, 96TP112-0, 1996.

T Dillon and D Niebur, *Neural Net Application in Power Systems*, CRL Publishing Ltd, 1996.

K P Wong and Y W Wong, 'Combined genetic algorithm/simulated annealing/fuzzy set approach to short-term generation scheduling with take-or-pay fuel contract', *IEEE Transactions on Power Systems*, **11**, 1996, 128-136.

Yann-Chang Huang, Hong-Tzer Yang and Ching-Lien Huang, 'Developing a new transformer fault diagnosis system, through evolutionary fuzzy logic', *IEEE Transactions on Power Delivery*, **12**, 1997, 761-767.

Z Michalewicz, *Genetic Algorithms + Data Structures = Evolution Programs*, 3rd edn. Springer-Verlag, Berlin,1996.

M. Kezunovic and I Rikalo, 'Detect and classify faults using neural nets', *IEEE Magazine on Computer Applications in Power*, October 1996, 42-47.

Z Michalewicz and M Schoenauer, 'Evolutionary algorithms for constrained parameter optimization problems', *Evolutionary Computation*, **4**, 1996, 1-32.

J M Render and S P Flasse, 'Hybrid methods using genetic algorithms for global optimization', *IEEE Transactions on Systems, Man and Cybernetics*, **26**, 1996, 243-258.

H Myung and J H Kim, 'Hybrid evolutionary programming for heavily constrained problems', *BioSystems*, **38**, 1996, 29-43.

J H Kim and H Myung, 'A two phase evolutionary programming for general constrained optimization problem,' *Proceedings of the Sixth Annual Conference on Evolutionary Programming*, L J Fogel, P J Angeline and T Bäck, Eds, MIT Press, Cambridge, MA, 1996, 295-304.

H-T Yang, P-C Yang and C-L Huang, 'Evolutionary programming based economic dispatch for units with non-smooth fuel cost functions', *IEEE Transactions on Power Systems*, **11**, 1996, 112-118.

M Mitchell, *An Introduction to Genetic Algorithms*, MIT Press, Cambridge, MA, 1996.

J M Renders and S P Flasse, 'Hybrid methods using genetic algorithms for global optimization', *IEEE Transactions on Systems, Man and Cybernetics*, **26**, 1996, 243-258.

T Bäck, *Evolutionary algorithms in theory and practice*, Oxford University Press, New York, 1996.

J R S Mantovani and A V Garcia, 'A heuristic method for reactive power planning', *IEEE Transactions on Power Systems*, **11**, 1996, 68-74.

H T Yang, P C Yang and C L Huang, 'Evolutionary programming based economic dispatch for units with non-smooth full cost function', *IEEE Transactions on Power Systems'*, **11**, 1996, 112-118.

J Smith and T C Fogarty, 'Self adaptation of mutation rates in a steady state genetic algorithm', *Proceedings of the Third IEEE Conference on Evolutionary Computation*, 1996, 318-323.

P J Angeline, 'The effects of noise on self-adaptive evolutionary optimization', *Proceedings of the Fifth Annual Conference on Evolutionary Programming*, MIT Press, Cambridge, MA, 1996,

T Bäck and H P Schwefel, 'Evolutionary computation: An overview', *Proceedings of the Third IEEE Conference on Evolutionary Computation*, 1996, 20-29.

J B Lee, J T Ma and L L Lai, 'Operation scheduling of multi-cogeneration using genetic algorithms', *Proceedings of the Power and Energy '96, IEE of Japan*, 1996, 93-98.

T Dalstein, T Friedrich, B Kulicke and D Sobajic, 'Multi neural network based fault area estimation for high speed protective relaying', *IEEE Transactions on Power Delivery*, **11**, 1996, 740-747.

Z Michalewicz and M Schoenauer, 'Evolutionary algorithms for constrained parameter optimization problems', *Evolutionary Computation*, **4**, 1996, 1-32.

Z Michalewicz, 'A survey of constraint handling techniques in evolutionary computation methods', *Proceedings of the Fourth Annual Conference on Evolutionary Programming*, J

R McDonnell, R G Reynolds and D B Fogel, Eds, MIT Press, Cambridge, MA, 1995, 135-155.

P H Chen and H C Chang, 'Large-scale economic dispatch', *IEEE Transactions on Power Systems,* **10**, 1995, 1919-1926.

M A El-Sharkawi, R J Marks II, Seho Oh and S J Huang, 'Localization of winding shorts using fuzzified neural networks', *IEEE Transactions on Energy Conversion,* **10**, 1995, 140-146.

Z Michalewicz, 'Genetic algorithms, numerical optimization, and constraints', *Proceedings of the Sixth International Conference on Genetic Algorithms,* L. J. Eshelman, Ed., Morgan Kaufmann, Los Altos, CA, 1995, 151-158.

Marimuthu Palaniswami, Yianni Attikiouzel, Robert J Marks II, David Fogel and Toshio Fukuda, Eds, *Computational Intelligence: a Dynamic System Perspective,* IEEE Press, 1995.

K Derr, *Applying OMT: a Practical Step-by-Step Guide to Using the Object Modelling Technique,* SIGS Books, 1995.

Vladimiro Miranda, J N Fidalgo, J A Pecas Lopes and L B Almeida, 'Real time preventive actions for transient stability enhancement with a hybrid neural network - optimization approach', *IEEE Transactions on Power Systems,* **10**, 1995, 1029-1035.

N M Roehl, C E Pedreira and H R Teles de Azevedo, 'Fuzzy ART neural network approach for incipient fault detection and isolation in rotating machines', *Proceedings of the International Conference on Neural Networks,* IEEE, Nov 1995, 538-542.

G B Sheble and K Brittig, 'Refined genetic algorithm - economic dispatch example', *IEEE Transactions on Power Systems,* **10**, 1995, 117-124.

D Srinivasan, C S Chang and A C Liew, 'Demand forecasting using fuzzy neural computation with special emphasis on weekend and public holiday forecasting', *IEEE Transactions on Power Systems,* **10**, 1995, 1897-1903.

D Papurt, *Inside the Object Model: the Sensible Use of C++,* SIGS Books, 1995.

A G Bakirtzis, J B Theocharis, S J Kiartzis and K J Satsois, 'Short term load forecasting using fuzzy neural networks', *IEEE Transactions on Power Systems,* **10**, 1995, 1518-1524.

P K Dash, G Ramakrishna, A C Liew and S Rahman, 'Fuzzy neural networks for time-series forecasting of electric load', *IEE Proceedings - Generation, Transmission and Distribution,* **142**, 1995, 535-544.

K Y Lee, X Bai and Y M Park, 'Optimization method for reactive power planning by using a modified simple genetic algorithm,' *IEEE Transactions in Power Systems,* **10**, 1995, 1843-1850.

Hong-Chan Chang and Mang-hui Wang, 'Neural network-based self-organizing fuzzy controller for transient stability of multimachine power systems', *IEEE Transactions on Energy Conversion,* **10**, June 1995, 339-347.

H P Schwefel, *Evolution and Optimum Seeking,* John Wiley & Sons, New York, 1995.

L J Eshelman and J D Schaffer, 'Productive recombination and propagating and preserving schemata,' *Foundations of Genetic Algorithms 3,* Morgan Kaufmann, San Francisco, CA, 1995, 299-313.

D B Fogel, *Evolution and Optimum Seeking,* Sixth-Generation Computer Technology Series, John Wiley & Sons, New York, 1995.

H G Beyer, 'Toward a theory of evolution strategies: self-adaptation', *Evolutionary Computation,* **3**, 1995, 311-348.

N Saravanan, D B Fogel, and K M Nelson, 'A comparison of methods for self-adaptation in evolutionary algorithms', *BioSystems,* **36**, 1995, 157-166.

W M Spears, 'Adapting crossover in evolutionary algorithms', *Proceedings of the Fourth Annual Conference on Evolutionary Programming,* MIT Press, Cambridge, MA, 1995, 367-384.

L D Whitley, 'Genetic algorithms and neural networks', *Genetic Algorithms in Engineering and Computer Science,* G Winter, J Periaux, M Galan and P Cuesta, Eds, John Wiley & Sons, Chichester, UK, 1995, 203-216.

C M Fonseca and P J Fleming, 'An overviw of evolutionary algorithms in multiobjective optimization', *Evolutionary Computation*, **3**, 1995, 1-16.

D Duffy, *'From Chaos to Classes: Object-Oriented Software Development in C++*, McGraw-Hill, 1995.

R Zitar and M Hassoun, 'Neurocontrollers trained with rules extracted by genetic assisted reinforcement learning system', *IEEE Transactions on Neural Networks*, **6**, 1995, 859-879.

Y Zhang, O P Malik, G S Hope and G P Chen, 'Application of an inverse input/output mapped ANN as a power system stabilizer', *IEEE Transactions on Energy Conversion*, **9**, 1994, 433-441.

Z Michalewicz and N Attia, 'Evolutionary optimization of constrained problems', *Proceedings of the Third Annual Conference on Evolutionary Programming*, A V Sebald and L J Fogel, Eds, World Scientific, River Edge, NJ, 1994, 98-108.

Z Michalewicz, 'Evolutionary operators for continuous convex parameters spaces', *Proceedings of the Third Annual Conference on Evolutionary Programming*, A. V. Sebald and L. J. Fogel, Eds, World Scientific, River Edge, NJ, 1994, 84-97.

A S Homaifar, H Y Lai, and X Qi, 'Constrained optimization via genetic algorithms', *Simulation*, **62**, 1994, 242-254.

R G Reynolds, 'An introduction to cultural algorithms', *Proceedings of the Third Annual Conference on Evolutionary Programming*, A V Sebald and L J Fogel, Eds, World Scientific, River Edge, NJ, 1994, 131-139.

N Saravanan and D B Fogel, 'Evolving neurocontrollers using evolutionary programming', *Proceedings of the First IEEE Conference on Evolutionary Computation*, Z Michalewicz. H Kitano, D Schal-fer, H P Schwefel and D B Fogel, Eds, IEEE Press, Piscataway. NJ, 1994, 217-222.

G Rudolph, 'Convergence analysis of canonical genetic algorithms', *IEEE Transactions on Neural Networks: Special Issue on Evolutionary Computation*, **5**, 1994, 96-101.

N Saravanan and D B Fogel, 'Evolving neurocontrollers using evolutionary programming', *Proceedings of the First IEEE Conference on Evolutionary Computation*, **1**, 1994, 217-222.

N Saravanan and D B Fogel, 'Learning of strategy parameters in evolutionary programming: an empirical study', *Proceedings of the Third Annual Conference on Evolutionary Programming*, World Scientific, Singapore, 1994, 269-280.

T Bäck, 'Selective pressure in evolutionary algorithms: a characterization of selection mechanisms', *Proceedings of the First IEEE Conference on Evolutionary Computation*, 1994, 57-62.

J C Bezek, 'What is computational intelligence?', *Computational Intelligence: Imitating Life*, J M Zurada, R J Marks II and C J Robinson, Eds, IEEE Press, New York, 1994, 1-12.

G Booch, *Object-Oriented Analysis and Design with Applications*, 2nd edn, Benjamin/Cummings, 1994.

P J Werbos, *The Roots of Backpropagation: From Ordered Derivatives to Neural Nets and Political Forecasting*, John Wiley & Sons, 1994.

E Cox, *The Fuzzy Systems Handbook: a Practitioner's Guide to Building, Using and Maintaining Fuzzy Systems*, Academic Press, 1994.

L G Perez, A J Flechsig, J L Meador and Z Obradovic, 'Training an ANN to discriminate between magnetising inrush and internal faults', *IEEE Transactions on Energy Conversion*, **9**, 1994, 434-441.

A D Papalexopoulos, S Hao and T M Peng, 'An implementation of a neural network based load forecasting model for the EMS', *IEEE Transactions on Power Systems*, **9**, 1994, 1956-1962.

Hisao Ischibuchi and Ryosuke Fujioka, 'Neural networks that learn from fuzzy IF-THEN rules', *IEEE Transactions on Fuzzy Systems*, **1**, 1993, 85-97.

D Powell and M M Skolnick, 'Using genetic algorithms in engineering design optimization with nonlinear constraints', *Proceedings of the Fifth International Conference on Genetic Algorithms*, S Forrest, Ed., Morgan Kaufmann, Los Altos, CA, 1993, 424-430.

M Schoenauer and S Xanthakis, 'Constrained GA optimization', *Proceedings of the Fifth International Conference on Genetic Algorithms*, S Forrest, Ed., Morgan Kaufmann, Los Altos, CA, 1993, 573-580.

S Forrest and M Mitchell, 'What makes a problem hard for a genetic algorithm? Some anomalous results and their explanation,' *Machine Learning*, **13**, 1993, 285-319.

M Mandischer, 'Representation and evolution of neural networks', *Artificial Neural Network and Genetic Algorithms*, R F Albrecht, C R Reeves and N C Steele, Eds, Springer-Verlag, Wien, Germany, 1993, 643-649.

D B Fogel, 'On the philosophical differences between evolutionary algorithms and genetic algorithms', *Proceedings of the Second Annual Conference on Evolutionary Programming*, Evolutionary Programming Society, San Diego, CA, 1993, 23-29.

W Wienholt, 'Minimizing the system error in feedforward neural networks with evolutionary srategy', *Proceedings of the International Conference on Artificial Neural Networks*, S Gilen and B Kappen, Eds, Springer, London, 1993, 490-493.

S B Haffner and A V Sebald, 'Computer-aided design of fuzzy HVAC controllers using evolutionary programming', *Proceedings of the Second Annual Conference on Evolutionary Programming*, Evolutionary Programming Society, San Diego, CA, 1993, 98-107.

M Dorigo and V Maniezzo, 'Parallel genetic algorithms: introduction and overview of current research', *Parallel Genetic Algorithms: Theory & Applications, Frontiers in Artificial Intelligence and Applications*, J Stender, Ed., IOS, Amsterdam, The Netherlands: 1993, 5-42.

D Hush and B Horne, 'Progress in unsupervised learning', *IEEE Signal Processing Magazine*, **10**, 1993, 8-39.

F Ndeh-Che, L L Lai and K Chu, 'The design of neural networks with object-oriented techniques', *IEE Colloquium on Recent Progress in Object Technology*, 1993.

L L Lai and F Ndeh-Che, 'An application of neural networks to improving power system stability', *IEE Colloquium on Advances in Neural Networks for Control System*, 1993.

J R McDonnell and D Waagen, 'Neural network structure design by evolutionary programming', *Proceedings of the Second Annual Conference on Evolutionary Programming*, Evolutionary Programming Society, La Jolla, CA, 1993, 79-89.

R Jang, 'ANFIS: adaptive-network-based fuzzy inference system', *IEEE Transactions on Systems, Man and Cybernetics*, **23**, 1993, 665-685.

R E Wilson and J M Nordstrom, 'EMTP transient modelling of a distance relay and comparison with EMTP laboratory testing' *IEEE Transaction on Power Delivery*, **8**, 1993, 984-992.

J Nie and D A Linkens, 'Learning control using fuzzified self-organizing radial basis function network,' *IEEE Transactions on Fuzzy Systems*, **1**, 1993, 280-287.

S Abe, 'Global convergence and suppression of spurious states of the Hopfield neural networks', *IEEE Transactions on Circuits and Systems*, **40**, 1993.

H Ischibuchi and R Fujioka, 'Neural networks that learn from fuzzy IF-THEN rules', *IEEE Transactions on Fuzzy Systems*, **1**, 1993, 85-97.

'Artificial neural networks for power systems: a literature survey', *CIGRE TF 38-06-06*, **1**, Convener: D Niebur, Dec 1993.

D Waagen, P Dercks and J McDonnell, 'The stochastic direction set algorithm: a hybrid technique for finding function extrema', *Proceedings of the First Annual Conference on Evolutionary Programming*, Evolutionary Programming Society, La Jolla, CA, 1992, 35-42.

H Mori, Y Tamaru, and S Tsuzuki, 'An artificial neural-net based technique for power system dynamic stability with the Kohonen model', *IEEE Transactions on Power Systems*, **7**, 1992, 856-864.

S Horikawa, T Furuhashi and Y Uchikawa, 'On fuzzy modelling using fuzzy neural networks with the back propagation algorithm', *IEEE Transactions on Neural Networks*, **3**, 1992, 801-806.

A F Sultan, G W Swift and D J Federchuk, 'Detection of high impedance arcing faults using a multi layer perceptron', *IEEE Transaction on Power Delivery*, **7**, October 1992, 1871-1877.

Y Ma and M A Shanblatt, 'A two-phase optimization neural network', *IEEE Transactions on Neural Networks*, **3**, 1992, 1003-1009.

S Zhang, X Zhu, and L H Zou, 'Second-order neural nets for constrained optimization', *IEEE Transactions on Circuits and Systems*, **39**, 1992, 1021-1024.

S Zhang and A G Constantinides, 'Lagrange programming neural networks', *IEEE Tranactions on Circuits and Sys*tems, **39**, 1992, 441-452.

K A De Jong, 'Are genetic algorithms function optimizers?', *Parallel Problem Solving from Nature 2*. Elsevier, Amsterdam, The Netherlands, 1992, 3-13.

D E Goldberg, K Deb and J H Clark, 'Genetic algorithms, noise, and the sizing of populations', *Complex Systems*, **6**,1992, 333-362.

D B Fogel, 'Evolving artificial intelligence,' *PhD dissertation*, University of California, San Diego, 1992.

M Herdy, 'Reproductive isolation as strategy parameter in hierarchically organized evolution strategies', *Parallel Problem Solving from Nature 2*, Elsevier, Amsterdam, The Netherlands, 1992, 207-217.

A Blum, *Neural networks in C++: an Object-Oriented Framework for Building Connectionist Systems,* John Wiley & Sons, 1992.

Y Takefuji, *Neural Network Parallel Computing*, Kluwer, 1992, 157-176.

Z Michalewicz and C Janikow, 'Handling constraints in genetic algorithms', *Proceedings of the Fourth International Conference on Genetic Algorithms*, R. Belew and L. Booker, Eds, Morgan Kaufmann, Los Altos, CA, 1991, 151-157.

R K Belew, J McInerney and N N Schraudolph, 'Evolving networks: using the genetic algorithm with connectionist learning', *Artificial Life II*, C G Langton, C Taylor, J D Farmer and S Rasmussen, Eds, Addison-Wesley, Reading, MA, 1991, 511-547.

C Z Janikow and Z Michalewicz, 'An experimental comparison of binary and floating point representations in genetic algorithms', *Proceedings of the Fourth International Conference on Genetic Algorithms*, Morgan Kaufmann, San Mateo, CA, 1991, 31-36.

C L Karr, 'Genetic algorithms for fuzzy controllers', *AI Expert*, **6**, 1991, 27-33.

P Thrift, 'Fuzzy logic synthesis with genetic algorithms', *Proceedings of the Fourth International Conference on Genetic Algorithms*, Morgan Kaufmann, San Mateo, CA, 1991, 514-518.

C Bishop, 'Improving the generalisation properties of radial basis function networks', *Neural Computation*, **3**, MIT, 1991, 579-588.

Y Y Hsu and C Yang, 'Design of artificial neural networks for short term load forecasting, part 1: self-organising feature maps for day type identification', *IEE Proceedings-C*, **138**, 1991, 407-413.

Y Y Hsu and C Yang, 'Design of artificial neural networks for short term load forecasting, part 2: multilayer feedforward networks for peak load and valley load forecasting', *IEE, Proceedings-C*, **138**, 1991, 407-413.

M Aggoune, M A El-Sharkawi, D C Park, M J Damborg and R J Marks, 'Preliminary results using artificial neural networks for security assessment', *IEEE Transactions on Power Systems*, **6**, 1991, 890-896.

B Kosko, *Neural Networks and Fuzzy Systems: A Dynamical Approach to Machine Intelligence*, Prentice-Hall, Englewood Cliffs, NJ, 1991.

C T Lin and C S G Lee, 'Neural network-based fuzzy logic control and decision system', *IEEE Transactions on Computers*, **40**, 1991, 1320-1336.

R Hetcht-Nelson, *Neurocomputing*, Addison-Wesley, 1990.

H Kitano, 'Designing neural networks using genetic algorithms with graph generation system', *Complex Systems*, **4**, 1990.

S Ebron, D Lubkeman and M White, 'A neural network approach to the detection of incipient faults in power distribution feeders', *IEEE Transactions on Power Delivery*, **5**, 1990, 905-914.

D B Fogel, L J Fogel and V W Porto, 'Evolving neural networks', *Biological Cybernetics*, **63**, 1990, 487-493.

B Widrow and M A Lehr, '30 years of adaptive neural networks: perceptron, madline, and backpropagation', *Proceedings of IEEE*, **78**, 1990, 1415-1442.

R C Ebehart and R W Dobbins, Eds, *Neural Networks PC Tools: a Practical Guide*, Academic Press, San Diego, 1990.

T Kohonen, *Self-Organisation and Associative Memory*, Springer-Verlag, 1989.

G F Miller, M Todd and S U Hegde, 'Designing neural networks using genetic algorithms', *Proceedings of the Third International Conference on Genetic Algorithms*, Morgan Kaufmann, San Mateo, CA, 1989, 379-384.

S Ebron, D L Lubkeman and M White, 'A neural network approach to the detection of incipient faults on power distribution feeders', *IEEE Transactions on Power Delivery*, **5**, 1990, 905-914.

I Moghram and S Rahman, 'Analysis and evaluation of five short term load forecasting techniques', *IEEE Transactions on Power Systems*, **4**, 1989, 1484-1491.

Y H Pao, *Adaptive Pattern Recognition and Neural Networks*, Addison-Wesley, Reading, MA, 1989.

D B Fogel, 'An evolutionary approach to the traveling salesman problem', *Biological Cybernetics*, **60**, 1988, 139-144.

B Kosko, 'Bi-directional associative memories', *IEEE Transactions on Systems, Man and Cybernetics*, **SMC-L8**, 1988, 49-60.

M Sugeno and G T Kang, 'Structure identification of fuzzy model', *Fuzzy Sets and Systems*, **28**, 1988, 15-53.

R Lippman, 'Introduction to computing with neural networks', *IEEE ASSP Magazine*, 1987, 4-20.

J J Grefenstette, 'Optimization of control parameters for genetic algorithms', *IEEE Transactions on Systems, Man and Cybernetics*, **SMC-16**, 1986, 122-128.

T Takagi and M Sugeno, 'Fuzzy identification of system and its application to modelling and control', *IEEE Transactions on Systems, Man and Cybernetics*, **15**, 1985, 116-132.

K Fukushima, S Miyake and T Ito, 'Neocognitron: a neural network model for a mechanism of visual pattern recognition', *IEEE Transactions on Systems, Man and Cybernetics*, 1983, 826-834.

J Hopfield, 'Neural networks and physical systems with emergent collective computational abilities', *Proceedings of the National Academy of Sciences*, USA, **79**, 1982, 2554-2558.

G H Burgin, 'Systems identification by quasilinearization and evolutionary programming', *Journal of Cybernetics*, **3**, 1973, 56-75.

T Kohonen, 'Correlation matrix memories', *IEEE Transactions on Computers*, **21**, 1972, 353-358.

L J Fogel, A J Owens and M J Walsh, *Artificial Intelligence Through Simulated Evolution*. John Wiley & Sons, New York, 1966.

L J Fogel, 'On the organization of intellect', *PhD dissertation*, University of California, Los Angeles, 1964.

H J Bremermann, 'Optimization through evolution and recombination', *Self-Organizing Systems*, M C Yovitsetal, Ed., Spartan, Washington, DC, 1962.

R M Friedberg, 'A learning machine: Part 1', *IBM Journal*, **2**, 1958, 2-13.

R M Friedberg, B Dunham and J H North, 'A learning machine: Part 2', *IBM Journal*, **3**, 1957, 282-287.

W S McCullogh and W Pitts, 'A logical calculus of the ideas imminent in logica calculus', *Bulletin of Mathematical Biophysics*, **5**, 1943, 115-133.

R A Fisher, *The Genetical Theory of Natural Selection*, Clarendon, Oxford , UK, 1930.

INDEX

Printed and bound by CPI Group (UK) Ltd, Croydon, CR0 4YY

27/10/2024

14580295-0003